STUDENT SOLUTIONS MANUAL

Thomas Engel
University of Washington

QUANTUM CHEMISTRY
& SPECTROSCOPY

SECOND EDITION

Thomas Engel

Prentice Hall
New York Boston San Francisco
London Toronto Sydney Tokyo Singapore Madrid
Mexico City Munich Paris Cape Town Hong Kong Montreal

Acquisitions Editor: Dan Kaveney
Editor in Chief, Chemistry and Geosciences: Nicole Folchetti
Marketing Manager: Erin Gardner
Assistant Editors: Jessica Neumann and Carol DuPont
Managing Editor, Chemistry and Geosciences: Gina M. Cheselka
Project Manager: Traci Douglas
Operations Specialist: Maura Zaldivar
Supplement Cover Manager: Paul Gourhan
Supplement Cover Designer: Tina Krivoshein
Cover Credit: Corbis/Superstock

© 2010 Pearson Education, Inc.

Pearson Prentice Hall

Pearson Education, Inc.

Upper Saddle River, NJ 07458

Pearson Prentice Hall™ is a trademark of Pearson Education, Inc.

The author and publisher of this book have used their best efforts in preparing this book. These efforts include the development, research, and testing of the theories and programs to determine their effectiveness. The author and publisher make no warranty of any kind, expressed or implied, with regard to these programs or the documentation contained in this book. The author and publisher shall not be liable in any event for incidental or consequential damages in connection with, or arising out of, the furnishing, performance, or use of these programs.

Printed in the United States of America

V036   10   9   8   7   6   5   4   3   2

ISBN-13:   978-0-321-61618-0
ISBN-10:   0-321-61618-9

Prentice Hall
is an imprint of

www.pearsonhighered.com

Contents

*Chapters 14, 15, and 17 do not contain solutions in the Solutions Manual

Chapter 1: From Classical to Quantum Mechanics

P1.5) Calculate the highest possible energy of a photon that can be observed in the emission spectrum of H.

The highest energy photon corresponds to a transition from $n = \infty$ to $n = 1$.

$$\tilde{v} = 109677\left(\frac{1}{1} - \frac{1}{\infty^2}\right) = 109677 \text{ cm}^{-1}$$

$$E = hc\tilde{v} = 2.17871 \times 10^{-18} \text{ J}$$

P1.8) What speed does a N_2 molecule have if it has the same momentum as a photon of wavelength 180. nm?

$$p = \frac{h}{\lambda} = m_{H_2} v_{H_2}$$

$$v_{H_2} = \frac{h}{m_{H_2}\lambda} = \frac{6.626 \times 10^{-34} \text{ J s}}{28.02 \text{ amu} \times 1.661 \times 10^{-27} \text{ kg (amu)}^{-1} \times 180. \times 10^{-9} \text{ m}} = 0.0791 \text{ m s}^{-1}$$

P1.12) Electrons have been used to determine molecular structure by diffraction. Calculate the speed of an electron for which the wavelength is equal to a typical bond length, namely, 0.175 nm.

$$v = \frac{p}{m} = \frac{h}{m\lambda} = \frac{6.626 \times 10^{-34} \text{ J s}}{9.109 \times 10^{-31} \text{ kg} \times 0.175 \times 10^{-9} \text{ m}} = 4.16 \times 10^6 \text{ m s}^{-1}$$

P1.13) For a monatomic gas, one measure of the "average speed" of the atoms is the root mean square speed, $v_{rms} = \langle v^2 \rangle^{1/2} = \sqrt{3kT/m}$, in which m is the molecular mass and k is the Boltzmann constant. Using this formula, calculate the de Broglie wavelength for H and Ne atoms at 250. and at 750. K.

$$\lambda = \frac{h}{mv_{rms}} = \frac{h}{\sqrt{3kTm}} = \frac{6.626 \times 10^{-34} \text{ J s}}{\sqrt{3 \times 1.381 \times 10^{-23} \text{ J K}^{-1} \times 250.\text{K} \times 1.008 \text{amu} \times 1.661 \times 10^{-27} \text{kg amu}^{-1}}}$$

$$= 1.59 \times 10^{-10} \text{ m}$$

or H at 750. K. $\lambda = 9.19 \times 10^{-11}$ m for He. For Ne, $\lambda = 3.56 \times 10^{-11}$ m and 2.05×10^{-11} m at 250. K and 750. K, respectively.

P1.16) If an electron passes through an electrical potential difference of 1 V, it has an energy of 1 electron-volt. What potential difference must it pass through in order to have a wavelength of 0.225 nm?

$$E = \frac{1}{2}m_e v^2 = \frac{1}{2}m_e \times \left(\frac{h}{m_e \lambda}\right)^2 = \frac{h^2}{2m_e \lambda^2}$$

$$= \frac{\left(6.626 \times 10^{-34} \text{ J s}\right)^2}{2 \times 9.109 \times 10^{-31} \text{ kg} \times \left(2.25 \times 10^{-10} \text{ m}\right)^2} \times \frac{1\text{eV}}{1.602 \times 10^{-19} \text{ J}} = 29.7 \text{ eV}$$

The electron must pass through an electrical potential of 29.7 V.

P1.20) The power (energy per unit time) radiated by a black body per unit area of surface expressed in units of W m^{-2} is given by $P = \sigma T^4$ with $\sigma = 5.67 \times 10^{-8}$ W m^{-2} K^{-4}. The radius of the sun is 6.95×10^5 km and the surface temperature is 5750. K. Calculate the total energy radiated per second by the sun. Assume ideal blackbody behavior.

$$E = PA = \sigma T^4 \times 4\pi r^2$$

$$= 5.67 \times 10^{-8} \text{ W m}^{-2}\text{K}^{-4} \times \left(5750. \text{ K}\right)^4 \times 4\pi \times \left(6.95 \times 10^8 \text{ m}\right)^2$$

$$= 3.76 \times 10^{26} \text{ W}$$

P1.21) The work function of tungsten is 4.50 eV. What is the minimum frequency of light required to observe the photoelectric effect on W? If light with a 225-nm wavelength is absorbed by the surface, what is the velocity of the emitted electrons?

a) For electrons to be emitted, the photon energy must be greater than the work function of the surface.

$$E = h\nu \geq 4.50 \text{ eV} \times \frac{1.602 \times 10^{-19} \text{ J}}{\text{eV}} = 9.05 \times 10^{-19} \text{ J}$$

$$\nu \geq \frac{E}{h} \geq \frac{9.05 \times 10^{-19} \text{ J}}{6.626 \times 10^{-34} \text{ J s}} \geq 1.09 \times 10^{15} \text{ s}^{-1}$$

b) The outgoing electron must first surmount the barrier arising from the work function, so not all the photon energy is converted to kinetic energy.

$$E_e = h\nu - \phi = \frac{hc}{\lambda} - \phi$$

$$= \frac{6.626 \times 10^{-34} \text{ J s} \times 2.998 \times 10^8 \text{ m s}^{-1}}{225 \times 10^{-9} \text{ m}} - 7.21 \times 10^{-19} \text{ J} = 1.62 \times 10^{-19} \text{ J}$$

$$v = \sqrt{\frac{2E_e}{m_e}} = \sqrt{\frac{2 \times 1.62 \times 10^{-19} \text{ J}}{9.11 \times 10^{-31} \text{ kg}}} = 5.96 \times 10^5 \text{ m s}^{-1}$$

P1.23) Calculate the longest and the shortest wavelength observed in the Lyman series.

For the longest wavelength, the transition is from $n = 2$ to $n = 1$.

$$\tilde{v} = 109677\left(\frac{1}{1} - \frac{1}{2^2}\right) = 82257.8 \text{ cm}^{-1}$$

$$\lambda = \frac{1}{\tilde{v}} = 1.21569 \times 10^{-5} \text{ cm}$$

For the shortest wavelength, the transition is from $n = \infty$ to $n = 1$.

$$\tilde{v} = 109677\left(\frac{1}{1} - \frac{1}{\infty^2}\right) = 109677 \text{ cm}^{-1}$$

$$\lambda = \frac{1}{\tilde{v}} = 9.11768 \times 10^{-6} \text{ cm}$$

P1.24) A 1000-W gas discharge lamp emits 5.25 W of ultraviolet radiation in a narrow range centered near 325 nm. How many photons of this wavelength are emitted per second?

$$n' = \frac{E_{total}}{E_{photon}} = \frac{3.00 \text{ W} \times 1 \text{ J s}^{-1} \text{ W}^{-1}}{\dfrac{hc}{\lambda}} = \frac{5.25 \text{ W} \times 1 \text{ J s}^{-1} \text{ W}^{-1}}{\dfrac{6.626 \times 10^{-34} \text{ J s} \times 2.998 \times 10^8 \text{ m s}^{-1}}{325 \times 10^{-9} \text{ m}}} = 8.59 \times 10^{18} \text{ s}^{-1}$$

P1.27) Pulsed lasers are powerful sources of nearly monochromatic radiation. Lasers that emit photons in a pulse of 10-ns duration with a total energy in the pulse of 0.10 J at 1000 nm are commercially available.

a. What is the average power (energy per unit time) in units of watts (1 W = 1 J/s) associated with such a pulse?

b. How many 1000-nm photons are emitted in such a pulse?

a) $P = \dfrac{\Delta E}{\Delta t} = \dfrac{0.10 \text{ J}}{1.0 \times 10^{-8} \text{ s}} = 1.0 \times 10^7 \text{ J s}^{-1}$

b) $N = \dfrac{E_{pulse}}{E_{photon}} = \dfrac{E_{pulse}}{h\dfrac{c}{\lambda}} = \dfrac{0.10 \text{ J}}{6.626 \times 10^{-34} \text{ J s}^{-1} \times \dfrac{2.998 \times 10^8 \text{ m s}^{-1}}{1.000 \times 10^6 \text{ m}}} = 5.0 \times 10^{17}$

Chapter 2: The Schrödinger Equation

P2.1) A wave traveling in the z direction is described by the wave function $\Psi(z,t) = A_1 \mathbf{x} \sin(kz-\omega t + \emptyset_1) + A_2 \mathbf{y} \sin(kz-\omega t + \emptyset_2)$, where $\mathbf{x}$ and $\mathbf{y}$ are vectors of unit length along the x and y axes, respectively. Because the amplitude is perpendicular to the propagation direction, $\Psi(z,t)$ represents a transverse wave.

a) What requirements must A_1 and A_2 satisfy for a plane polarized wave in the x-z plane? The amplitude of a plane polarized wave is non-zero only in one plane

b) What requirements must A_1 and A_2 satisfy for a plane polarized wave in the y-z plane?

c) What requirements must A_1 and A_2 and ϕ_1 and ϕ_2 satisfy for a plane polarized wave in a plane oriented at 45° to the xz plane?

d) What requirements must A_1 and A_2 and ϕ_1 and ϕ_2 satisfy for a circularly polarized wave? The phases of the two components of a circularly polarized wave differ by $\pi/2$.

a) The amplitude along the x axis must oscillate, and the amplitude along the y axis must vanish. Therefore $A_1 \neq 0$ and $A_2 = 0$.

b) The amplitude along the y axis must oscillate, and the amplitude along the x axis must vanish. Therefore $A_1 = 0$ and $A_2 \neq 0$.

c) The amplitude along both the x and y axes must oscillate. Therefore $A_1 \neq 0$ and $A_2 \neq 0$. Because they must oscillate in phase, $\phi_1 = \phi_2$.

d) The amplitude along both the x and y axes must oscillate with the same amplitude. Therefore $A_1 = A_2 \neq 0$. For a circularly polarized wave, the x and y components must be out of phase by $\pi/2$.

Therefore $\phi_1 = \phi_2 \pm \dfrac{\pi}{2}$. This can be seen by comparing the x and y amplitudes for the positive sign.

$\Psi(z,t) = A_1 \mathbf{x} \sin(kz - \omega t + \phi_1) + A_1 \mathbf{y} \sin(kz - \omega t + \phi_1 + \dfrac{\pi}{2})$

let $kz + \phi = kz'$

$\Psi(z,t) = A_1 \mathbf{x} \sin(kz' - \omega t) + A_1 \mathbf{y} \sin(kz' - \omega t + \dfrac{\pi}{2})$

$\quad = A_1 \mathbf{x} \sin(kz' - \omega t) + A_1 \mathbf{y} \left[\sin(kz' - \omega t)\cos\dfrac{\pi}{2} + \cos(kz' - \omega t)\sin\dfrac{\pi}{2} \right]$

$\quad = A_1 \mathbf{x} \sin(kz' - \omega t) + A_1 \mathbf{y} \cos(kz' - \omega t)$

4

The x and y amplitudes are $\pi/2$ out of phase and the sum of the squares of their amplitudes is a constant as required for a circle.

P2.2) Because $\int_0^d \cos(n\pi x/d)\cos(m\pi x/d)dx = 0$, $m \neq n$, the functions $\cos(n\pi x/d)$ for $n = 1, 2, 3, ...$ form an orthogonal set. What constant must these functions be multiplied by to form an orthonormal set?

$$1 = N^2 \int_0^d \cos\left(\frac{m\pi x}{d}\right)\cos\left(\frac{m\pi x}{d}\right)dx = N^2\left[\frac{x}{2} + \frac{d}{4m\pi}\sin\left(\frac{2m\pi x}{d}\right)\right]_0^d$$

where we have used the standard integral $\int\left(\cos^2 ax\right)dx = \frac{1}{2}x + \frac{1}{4a}\sin 2ax$

$$1 = N^2\left[\frac{d}{2} + \frac{d}{4m\pi}\sin\left(2m\pi\right) - \frac{0}{2} - \frac{d}{4m\pi}\sin\left(0\right)\right] = \frac{d}{2}N^2 = 1$$

$$N = \sqrt{\frac{2}{d}}$$

P2.6) Carry out the following coordinate transformations:

a. Express the point $x = 4$, $y = 2$, and $z = 3$ in spherical coordinates.

b. Express the point $r = 7$, $\theta = \dfrac{\pi}{8}$, and $\phi = \dfrac{5\pi}{8}$ in Cartesian coordinates.

a) $r = \sqrt{x^2 + y^2 + z^2} = \sqrt{4^2 + 2^2 + 3^2} = \sqrt{29}$

$$\theta = \cos^{-1}\frac{z}{\sqrt{x^2 + y^2 + z^2}} = \cos^{-1}\frac{2}{\sqrt{29}} = 0.980 \text{ radians}$$

$$\phi = \tan^{-1}\frac{y}{x} = \tan^{-1}\frac{2}{4} = 0.464 \text{ radians}$$

b) $x = r\sin\theta\cos\phi = 7\sin\dfrac{\pi}{8}\cos\dfrac{5\pi}{8} = -1.03$

$$y = r\sin\theta\sin\phi = 7\sin\frac{\pi}{8}\sin\frac{5\pi}{8} = 2.75$$

$$z = r\cos\theta = 7\cos\frac{\pi}{8} = 6.47$$

P2.8) Show that

$$\frac{a+ib}{c+id} = \frac{ac+bd+i(bc-ad)}{c^2+d^2}$$

$$\frac{a+ib}{c+id} = \left(\frac{a+ib}{c+id}\right)\left(\frac{c-id}{c-id}\right) = \frac{ac+bd+ibc-iad}{c^2+d^2} = \frac{ac+bd+i(bc-ad)}{c^2+d^2}$$

P2.9) Express the following complex numbers in the form $re^{i\theta}$.

a. $7-3i$

c. $\dfrac{7-i}{5+3i}$

b. $-5i$

d. $\dfrac{4+i}{1-2i}$

In the notation $re^{i\theta}$, $r = |z| = \sqrt{a^2+b^2}$ and $\theta = \sin^{-1}\left(\dfrac{\text{Im } z}{|z|}\right)$.

a) $7-3i = \sqrt{58}\exp\left(i\sin^{-1}\dfrac{-3}{\sqrt{58}}\right) = \sqrt{58}\exp(-0.129i\pi)$

b) $-5i = 5\exp\left(i\sin^{-1}\dfrac{-5}{\sqrt{25}}\right) = 5\exp\left(-\dfrac{\pi}{2}\right)$

c)

$$\frac{7-i}{3+5i} = \frac{7-i}{3+5i} \times \frac{3-5i}{3-5i} = \frac{21-38i-5}{34} = \frac{8}{17} - \frac{19i}{17}$$

We next calculate the magnitude of the complex number

$$\sqrt{\left(\frac{8}{17}\right)^2 + \left(-\frac{19}{17}\right)^2} = \sqrt{\frac{425}{289}} = \sqrt{\frac{25}{17}} = \frac{5}{\sqrt{17}}$$

$$= \frac{5}{\sqrt{17}}\exp\left(i\sin^{-1}\frac{19/17}{5/\sqrt{17}}\right) = \frac{5}{\sqrt{17}}\exp(-0.217i\pi)$$

d)

$$\frac{4+i}{1-2i} = \frac{4+i}{1-2i} \times \frac{1+2i}{1+2i} = \frac{4+9i-2}{5} = \frac{2}{5} + \frac{9i}{5}$$

We next calculate the magnitude of the complex number

$$\sqrt{\left(\frac{2}{5}\right)^2 + \left(\frac{9}{5}\right)^2} = \sqrt{\frac{85}{25}} = \sqrt{\frac{17}{5}}$$

$$= \sqrt{\frac{17}{5}} \exp\left(i\sin^{-1}\frac{9/5}{\sqrt{17/5}}\right) = \sqrt{\frac{17}{5}} \exp(0.430i\pi)$$

P2.14) Determine in each of the following cases if the function in the first column is an eigenfunction of the operator in the second column. If so, what is the eigenvalue?

a. $3\cos^2\theta - 1$ $\qquad\qquad \frac{1}{\sin\theta}\frac{d}{d\theta}\left(\sin\theta\frac{d}{d\theta}\right)$

b. $e^{-(x^2/2)}$ $\qquad\qquad \frac{d^2}{dx^2} - x^2$

c. $e^{-4i\phi}$ $\qquad\qquad \frac{d^2}{d\phi^2}$

a)

$$\frac{1}{\sin\theta}\frac{d}{d\theta}\left(\sin\theta\frac{d\left(3\cos^2\theta - 1\right)}{d\theta}\right) = \frac{1}{\sin\theta}\frac{d}{d\theta}\left(-6\cos\theta\sin^2\theta\right)$$

$$= \frac{1}{\sin\theta}\left(6\sin^3\theta - 12\cos^2\theta\sin\theta\right) = 6\sin^2\theta - 12\cos^2\theta$$

$$= 6 - 18\cos^2\theta = -6\left(3\cos^2\theta - 1\right)$$

Eigenfunction with eigenvalue −6.

b)

$$\frac{d^2 e^{-\frac{1}{2}x^2}}{dx^2} - x^2 e^{-\frac{1}{2}x^2} = -e^{-\frac{1}{2}x^2}$$

Eigenfunction with eigenvalue −1.

c)

$$\frac{d^2 e^{-4i\phi}}{d\phi^2} = -16e^{-4i\phi}$$

Eigenfunction with eigenvalue –16.

P2.15) Show by carrying out the integration that $\sin(m\pi x/a)$ and $\cos(m\pi x/a)$, where m is an integer, are orthogonal over the interval $0 \le x \le a$. Would you get the same result if you used the interval $0 \le x \le 3a/4$? Explain your result..

$$\int_0^a \cos\left(\frac{m\pi x}{a}\right) \sin\left(\frac{m\pi x}{a}\right) dx = \left[\frac{a}{2m\pi}\sin^2\left(\frac{m\pi x}{a}\right)\right]_0^a = \frac{a}{2m\pi}\left[\sin^2(m\pi) - 0\right] = 0$$

$$\int_0^{\frac{3a}{4}} \cos\left(\frac{m\pi x}{a}\right) \sin\left(\frac{m\pi x}{a}\right) dx = \int_0^{\frac{3a}{4}} \cos\left(\frac{m\pi x}{a}\right) \sin\left(\frac{m\pi x}{a}\right) dx$$

$$= \left[\frac{a}{2m\pi}\sin^2\left(\frac{m\pi x}{a}\right)\right]_0^{\frac{3a}{4}} = \frac{a}{2m\pi}\left[\sin^2\left(\frac{3m\pi}{4}\right) - 0\right] \ne 0$$

except for the special case $\frac{3m}{4} = n$ where n is an integer. The length of the integration interval must be n periods (for n an integer) to make the integral zero.

P2.19) Is the function $3x^2 - 1$ an eigenfunction of the operator $-(1-x^2)(d^2/dx^2) + 2x(d/dx)$? If so, what is the eigenvalue?

$$-(1-x^2)(d^2[3x^2-1]/dx^2) + 2x(d[3x^2-1]/dx)$$
$$= -6(1-x^2) + 12x^2 = 18x^2 - 6 = 6[3x^2-1]$$

Eigenfunction with eigenvalue 6

P2.20) Find the result of operating with $d^2/dx^2 - 4x^2$ on the function e^{-ax^2}. What must the value of a be to make this function an eigenfunction of the operator?

$$\frac{d^2 e^{-ax^2}}{dx^2} - 4x^2 e^{-ax^2} = -2ae^{-ax^2} - 4x^2 e^{-ax^2} + 4a^2 x^2 e^{-ax^2} = -2ae^{-ax^2} + 4\left(a^2 - 1\right)x^2 e^{-ax^2}$$

For the function to be an eigenfunction of the operator, the terms containing $x^2 e^{-ax^2}$ must vanish. This is the case if $a = \pm 1$.

P2.21) Determine in each of the following cases if the function in the first column is an eigenfunction of the operator in the second column. If so, what is the eigenvalue?

a. $\sin\theta\cos\phi$ $\quad \partial/\partial\phi$

b. $e^{(-x^2/2)}$ $\quad (1/x)d/dx$

c. $\sin\theta$ $\quad (\sin\theta/\cos\theta)d/d\theta$

a)

$$\frac{\partial}{\partial\phi}\sin\theta\cos\phi = -\sin\theta\sin\phi. \qquad \text{Not an eigenfunction}$$

b)

$$\frac{1}{x}\frac{d}{dx}e^{-\frac{1}{2}x^2} = -e^{-\frac{1}{2}x^2} \qquad \text{Eigenfunction with eigenvalue } -1$$

c)

$$\frac{\sin\theta}{\cos\theta}\frac{d}{d\theta}\sin\theta = \sin\theta \qquad \text{Eigenfunction with eigenvalue } +1$$

P2.24) If two operators act on a wave function as indicated by $\hat{A}\hat{B}f(x)$, it is important to carry out the operations in succession with the first operation being that nearest to the function. Mathematically, $\hat{A}\hat{B}f(x) = \hat{A}(\hat{B}f(x))$ and $\hat{A}^2 f(x) = \hat{A}(\hat{A}f(x))$. Evaluate the following successive operations $\hat{A}\hat{B}f(x)$. The operators $\hat{A}$ and $\hat{B}$ are listed in the first and second columns and $f(x)$ is listed in the third column.

a. $\dfrac{d}{dx}$ $\qquad \dfrac{d}{dx}$ $\qquad x^2 + e^{ax^2}$

b. $\dfrac{\partial^2}{\partial y^2}$ $\qquad \dfrac{\partial}{\partial x}$ $\qquad \cos 3y \sin^2 x$

c. $\dfrac{\partial}{\partial\theta}$ $\qquad \dfrac{\partial^2}{\partial\phi^2}$ $\qquad \dfrac{\cos\phi}{\sin\theta}$

9

a) $\dfrac{d}{dx}\left[\dfrac{d\left(x^2+e^{ax^2}\right)}{dx}\right]=\dfrac{d}{dx}\left[2x+2a\,x\,e^{ax^2}\right]=2+4a^2x^2e^{ax^2}+2a\,e^{ax^2}$

b) $\dfrac{\partial^2}{\partial y^2}\left[\dfrac{\partial\left(\cos 3y\sin^2 x\right)}{\partial x}\right]=\dfrac{\partial^2}{\partial y^2}\left[2\cos 3y\sin x\cos x\right]=-18\cos 3y\sin x\cos x$

c) $\dfrac{\partial}{\partial\theta}\left[\dfrac{\partial^2\left(\dfrac{\cos\phi}{\sin\theta}\right)}{\partial\phi^2}\right]=\dfrac{\partial}{\partial\theta}\left[-\dfrac{\cos\phi}{\sin\theta}\right]=\dfrac{\cos\phi\cos\theta}{\sin^2\theta}$

P2.26) Consider a two-level system with $\varepsilon_1 = 3.10\times10^{-21}$ J and $\varepsilon_2 = 6.10\times10^{-21}$ J. If $g_2 = g_1$, what value of T is required to obtain $n_2/n_1 = 0.225$? What value of T is required to obtain $n_2/n_1 = 0.875$?

$$\frac{n_2}{n_1}=\frac{g_2}{g_1}\exp\left[\frac{-\left(\varepsilon_2-\varepsilon_1\right)}{kT}\right]$$

$$\ln\left(\frac{n_2}{n_1}\right)=\ln\left(\frac{g_2}{g_1}\right)-\frac{\left(\varepsilon_2-\varepsilon_1\right)}{kT}$$

$$\frac{1}{T}=\frac{k}{\left(\varepsilon_2-\varepsilon_1\right)}\left[\ln\left(\frac{g_2}{g_1}\right)-\ln\left(\frac{n_2}{n_1}\right)\right]$$

$$T=\frac{\left(\varepsilon_2-\varepsilon_1\right)}{k\left[\ln\left(\dfrac{g_2}{g_1}\right)-\ln\left(\dfrac{n_2}{n_1}\right)\right]}$$

for $n_2/n_1 = 0.225$ $T=\dfrac{3.00\times10^{-21}\text{ J}}{1.381\times10^{-23}\text{ J K}^{-1}\times\left[\ln(1)-\ln(0.225)\right]}=145$ K

for $n_2/n_1 = 0.875$ $T=\dfrac{3.00\times10^{-21}\text{J}}{1.381\times10^{-23}\text{J K}^{-1}\times\left[\ln(1)-\ln(0.875)\right]}=1.63\times10^{3}$ K

P2.28) Normalize the set of functions $\phi_n(\theta)=e^{in\theta}$, $0\le\theta\le2\pi$. To do so, you need to multiply the functions by a so-called normalization constant N so that the integral

$N\,N^*\int_0^{2\pi}\phi_m^*(\theta)\phi_n(\theta)d\theta=1$ for $m=n$.

$$N N^* \int_0^{2\pi} e^{-in\theta} e^{in\theta} d\theta = N N^* \int_0^{2\pi} d\theta = 2\pi N N^* = 1$$ This is satisfied for $N = \dfrac{1}{\sqrt{2\pi}}$ and the normalized

functions are $\phi_n(\theta) = \dfrac{1}{\sqrt{2\pi}} e^{in\theta}, \quad 0 \le \theta \le 2\pi.$

P2.30) Operate with a) $\dfrac{\partial}{\partial x} + \dfrac{\partial}{\partial y} + \dfrac{\partial}{\partial z}$ and b) $\dfrac{\partial^2}{\partial x^2} + \dfrac{\partial^2}{\partial y^2} + \dfrac{\partial^2}{\partial z^2}$ on the function

$A\cos(k_1 x)\cos(k_2 y)\cos(k_3 z)$. Is the function an eigenfunction of either operator? If so, what is the

eigenvalue?

$$\left[\frac{\partial}{\partial x} + \frac{\partial}{\partial y} + \frac{\partial}{\partial z} \right] A \cos k_1 x \cos k_2 y \cos k_3 z$$

$$= -Ak_1 \sin k_1 x \cos k_2 y \cos k_3 z - Ak_2 \cos k_1 x \sin k_2 y \cos k_3 z - Ak_3 \cos k_1 x \cos k_2 y \sin k_3 z$$

$$= -A\left[k_1 \sin k_1 x \cos k_2 y \cos k_3 z + k_2 \cos k_1 x \sin k_2 y \cos k_3 z + k_3 \cos k_1 x \cos k_2 y \sin k_3 z \right]$$

not an eigenfunction

$$\left[\frac{\partial^2}{\partial x^2} + \frac{\partial^2}{\partial y^2} + \frac{\partial^2}{\partial z^2} \right] A \cos k_1 x \cos k_2 y \cos k_3 z$$

$$= -A\left[k_1^2 + k_2^2 + k_3^2 \right] \cos k_1 x \cos k_2 y \cos k_3 z$$

eigenfunction with eigenvalue $-\left[k_1^2 + k_2^2 + k_3^2 \right]$

P2.36) Which of the following wave functions are eigenfunctions of the operator d/dx? If they are

eigenfunctions, what is the eigenvalue?

a. $ae^{-3x} + be^{-3ix}$ d. $\cos ax$

b. $\sin^2 x$ e. e^{-ix^2}

c. e^{-ix}

a) $\dfrac{d\left(ae^{-3x} + be^{-3ix} \right)}{dx} = -3ae^{-3x} - 3ibe^{-3ix}$ Not an eigenfunction

b) $\dfrac{d\sin^2 x}{dx} = 2\sin x \cos x$ Not an eigenfunction

c) $\dfrac{de^{-ix}}{dx} = -ie^{-ix}$ Eigenfunction with eigenvalue $-i$

d) $\dfrac{d \cos a x}{d x} = -a \sin a x$ Not an eigenfunction

e) $\dfrac{d\, e^{-i x^2}}{d x} = -2i\, x\, e^{-i x^2}$ Not an eigenfunction

Chapter 4: Using Quantum Mechanics on Simple Systems

P4.3) Normalize the total energy eigenfunctions for the three-dimensional box in the interval $0 \le x \le a, 0 \le y \le b, 0 \le z \le c$.

$$1 = \int_0^a \int_0^b \int_0^c \psi^*(x,y,z)\psi(x,y,z)\,dx\,dy\,dz$$

$$= N^2 \int_0^a \sin^2\left(\frac{n_x \pi x}{a}\right)dx \int_0^b \sin^2\left(\frac{n_y \pi y}{b}\right)dy \int_0^c \sin^2\left(\frac{n_z \pi z}{c}\right)dz$$

Using the standard integral $\int \sin^2 \alpha x\,dx = \dfrac{x}{2} - \dfrac{\sin(2\alpha x)}{4\alpha}$

$$1 = N^2 \int_0^a \sin^2\left(\frac{n_x \pi x}{a}\right)dx \int_0^b \sin^2\left(\frac{n_y \pi y}{b}\right)dy \int_0^c \sin^2\left(\frac{n_z \pi z}{c}\right)dz$$

$$1 = N^2 \left[\frac{a}{2} - \frac{a}{4n_x \pi}(\sin n_x \pi - \sin 0)\right] \times \left[\frac{b}{2} - \frac{b}{4n_y \pi}(\sin n_y \pi - \sin 0)\right] \times \left[\frac{c}{2} - \frac{c}{4n_z \pi}(\sin n_z \pi - \sin 0)\right] = N^2 \frac{abc}{8}$$

$$N = \sqrt{\frac{8}{abc}} \quad \text{and} \quad \psi(x,y) = \sqrt{\frac{8}{abc}} \sin\left(\frac{n_x \pi x}{a}\right) \sin\left(\frac{n_y \pi y}{b}\right) \sin\left(\frac{n_z \pi z}{c}\right)$$

P4.4) Is the superposition wave function for the free particle an eigenfunction of the momentum operator? Is it an eigenfunction of the total energy operator? Explain your result.

$$-i\hbar \frac{d}{dx}\left(A_+ e^{+i\sqrt{\frac{2mE}{\hbar^2}}x} + A_- e^{-i\sqrt{\frac{2mE}{\hbar^2}}x}\right) = -(i)^2 \hbar \sqrt{\frac{2mE}{\hbar^2}} A_+ e^{+i\sqrt{\frac{2mE}{\hbar^2}}x} + (i)^2 \hbar \sqrt{\frac{2mE}{\hbar^2}} A_- e^{-i\sqrt{\frac{2mE}{\hbar^2}}x}$$

$$= \hbar k\, A_+ e^{+i\sqrt{\frac{2mE}{\hbar^2}}x} - \hbar k\, A_- e^{-i\sqrt{\frac{2mE}{\hbar^2}}x}$$

This function is not an eigenfunction of the momentum operator, because the operation does not return the original function multiplied by a constant.

$$-\frac{\hbar^2}{2m}\frac{d^2}{dx^2}\left(A_+ e^{+i\sqrt{\frac{2mE}{\hbar^2}}x} + A_- e^{-i\sqrt{\frac{2mE}{\hbar^2}}x}\right) = -(i)^2 \frac{\hbar^2}{2m}\frac{2mE}{\hbar^2} A_+ e^{+i\sqrt{\frac{2mE}{\hbar^2}}x} - (-i)^2 \frac{\hbar^2}{2m}\frac{2mE}{\hbar^2} A_- e^{-i\sqrt{\frac{2mE}{\hbar^2}}x}$$

$$= E\left(A_+ e^{+i\sqrt{\frac{2mE}{\hbar^2}}x} + A_- e^{-i\sqrt{\frac{2mE}{\hbar^2}}x}\right)$$

This function is an eigenfunction of the total energy operator. Because the energy is proportional to p^2, the difference in sign of the momentum of these two components does not affect the energy.

P4.8) Evaluate the normalization integral for the eigenfunctions of $\hat{H}$ for the particle in the box $\psi_n(x) = A\sin(n\pi x/a)$ using the trigonometric identity $\sin^2 y = (1 - \cos 2y)/2$.

$$1 = \int_0^a A^2 \sin^2\left(\frac{n\pi x}{a}\right) dx$$

let $y = \dfrac{n\pi x}{a}; \; dx = \dfrac{a}{n\pi} dy$

$$1 = A^2 \frac{a}{n\pi} \int_0^{n\pi} \sin^2 y \, dy = A^2 \frac{a}{n\pi} \int_0^{n\pi} \frac{1 - \cos 2y}{2} dy = A^2 \frac{a}{n\pi} \left[\frac{y}{2} - \frac{\sin 2y}{4}\right]_0^{n\pi}$$

$$= \frac{A^2}{2} \frac{a}{n\pi} n\pi - \frac{A^2}{4} \frac{a}{n\pi}(\sin 2n\pi - \sin 0) = \frac{A^2 a}{2}$$

$$A = \sqrt{\frac{2}{a}}$$

P4.10) What is the solution of the time-dependent Schrödinger equation $\Psi(x,t)$ for the total energy eigenfunction $\psi_4(x) = \sqrt{2/a}\sin(4\pi x/a)$ in the particle in the box model? Write $\omega = E/\hbar$ explicitly in terms of the parameters of the problem.

$$\psi(x,t) = \psi(x)e^{-i\omega t} = \psi(x)e^{-i\frac{Et}{\hbar}}$$

Because $E = \dfrac{n^2 h^2}{8ma^2} = \dfrac{16 h^2}{8ma^2}$,

$$\psi(x,t) = \sqrt{\frac{2}{a}}\sin\left(\frac{4\pi x}{a}\right)e^{-i\frac{4\pi ht}{ma^2}}$$

P4.13) Show that the energy eigenvalues for the free particle, $E = \hbar^2 k^2/2m$, are consistent with the classical result $E = (1/2)mv^2$.

$$E = \frac{1}{2}mv^2 = \frac{p^2}{2m}$$

From the de Broglie relation, $p = \frac{h}{\lambda}$

$$E = \frac{1}{2m}\left(\frac{h}{\lambda}\right)^2 = \frac{\hbar^2 k^2}{2m},$$ showing consistency between the classical and quantum result.

P4.15) Calculate the wavelength of the light emitted when an electron in a one-dimensional box of length 3.0 nm makes a transition from the $n = 5$ state to the $n = 4$ state.

$$E = h\nu = \frac{h^2}{8ma^2}\left(n_2^2 - n_1^2\right) \quad \nu = \frac{h}{8ma^2}\left(n_2^2 - n_1^2\right)$$

$$= \frac{6.26 \times 10^{-34} \text{ J s}}{8 \times 9.109 \times 10^{-31} \text{ kg} \times 9.0 \times 10^{-18} \text{ m}^2}\left(5^2 - 4^2\right) = 9.1 \times 10^{13} \text{ s}^{-1}$$

$$\lambda = \frac{c}{\nu} = \frac{2.998 \times 10^8 \text{ m s}^{-1}}{9.1 \times 10^{13} \text{ s}^{-1}} = 3.3 \times 10^{-6} \text{ m}$$

P4.20) Calculate a) the zero point energy of a CO molecule in a one-dimensional box of length 1.00 cm and b) the ratio of the zero point energy to kT at 300.K.

$$E_1 = \frac{h^2}{8ma^2} = \frac{\left(6.26 \times 10^{-34} \text{ J s}\right)^2}{8 \times 28.01 \text{ amu} \times 1.661 \times 10^{-27} \text{ kg amu}^{-1} \times 1.00 \times 10^{-4} \text{ m}^2}$$

$$= 1.18 \times 10^{-37} \text{ J}$$

$$\frac{E_1}{kT} = \frac{1.18 \times 10^{-37} \text{ J}}{1.381 \times 10^{-23} \text{ J K}^{-1} \times 300. \text{ K}} = 2.85 \times 10^{-18}$$

P4.21) Normalize the total energy eigenfunction for the rectangular two-dimensional box, in the interval $0 \le x \le a, 0 \le y \le b$.

$$\psi_{n_x, n_y}(x, y) = N \sin\left(\frac{n_x \pi x}{a}\right) \sin\left(\frac{n_y \pi y}{b}\right)$$

$$1 = \int_0^a \int_0^b \psi^*(x,y)\psi(x,y)\,dx\,dy = N^2 \int_0^a \sin^2\left(\frac{n_x \pi x}{a}\right)dx \int_0^b \sin^2\left(\frac{n_y \pi y}{b}\right)dy$$

Using the standard integral $\int \sin^2(\alpha x)\,dx = \dfrac{x}{2} - \dfrac{\sin(2\alpha x)}{4\alpha}$

$$N^2 \int_0^a \sin^2\left(\frac{n_x \pi x}{a}\right)dx \int_0^b \sin^2\left(\frac{n_y \pi y}{b}\right)dy$$

$$= N^2\left[\frac{a}{2} - \frac{a}{4n_x \pi}(\sin n_x \pi - \sin 0)\right] \times \left[\frac{b}{2} - \frac{b}{4n_y \pi}(\sin n_y \pi - \sin 0)\right] = N^2\frac{ab}{4}$$

$$N = \sqrt{\frac{4}{ab}} \text{ and } \psi(x,y) = \sqrt{\frac{4}{ab}}\sin\left(\frac{n_x \pi x}{a}\right)\sin\left(\frac{n_y \pi y}{b}\right)$$

P4.24) What are the energies of the lowest 5 energy levels in a three-dimensional box with $a = b = c$?
What is the degeneracy of each level?

$n_x^2 + n_y^2 + n_z^2 = 3$: (111) degeneracy 1

$n_x^2 + n_y^2 + n_z^2 = 6$: (112), (121), (211) degeneracy 3

$n_x^2 + n_y^2 + n_z^2 = 9$: (212), (122), (221) degeneracy 3

$n_x^2 + n_y^2 + n_z^2 = 11$: (311), (131), (113) degeneracy 3

$n_x^2 + n_y^2 + n_z^2 = 12$: (222) degeneracy 1

P4.25) In discussing the Boltzmann distribution in Chapter 13, we used the symbols g_i and g_j to indicate the degeneracies of the energy levels i and j. By degeneracy, we mean the number of distinct quantum states (different quantum numbers), all of which have the same energy.

a. Using your answer to Problem P4.19a, what is the degeneracy of the energy level $5h^2/8ma^2$ for the square two-dimensional box of edge length a?

b. Using your answer to Problem P4.14b, what is the degeneracy of the energy level $9h^2/8ma^2$ for a three-dimensional cubic box of edge length a?

a) The only pairs n_x, n_y that satisfy the equation $n_x^2 + n_y^2 = 5$ are 2, 1 and 1, 2. Therefore the degeneracy of this energy level is 2.

b) The only trios of nonzero numbers n, q, r that satisfy the equation $n_x^2 + n_y^2 + n_z^2 = 9$ are 2, 2, 1; 2, 1, 2; and 1, 2, 2. Therefore the degeneracy of this energy level is 3.

P4.27) Two wave functions are distinguishable if they lead to a different probability density. Which of the following wave functions are distinguishable from $\sin kx$?

a. $\left(e^{ikx} - e^{-ikx}\right)/2$

b. $e^{i\theta} \sin kx$, θ a constant

c. $\cos\left(kx - \pi/2\right)$

d. $i\cos\left(kx + \pi/2\right)\left(\sin\theta + i\cos\theta\right)\left(-\dfrac{\sqrt{2}}{2} + i\dfrac{\sqrt{2}}{2}\right)$

Two wave functions ψ_1 and ψ_2 are indistinguishable if $\psi_1^*\psi_1 = \psi_2^*\psi_2$. For the wave function in the problem, $\left(\sin kx\right)\left(\sin kx\right) = \sin^2 kx$.

a)

$$\frac{\left(e^{ikx} - e^{-ikx}\right)\left(e^{-ikx} - e^{ikx}\right)}{2i \qquad\qquad 2i} = \frac{-i^2\left(\cos kx + i\sin kx - \left[\cos kx - i\sin kx\right]\right)^2}{4}$$

$$= \left(2\sin kx\right)^2/4 = \sin^2 kx \quad \text{indistinguishable}$$

b) $\left(e^{i\theta} \sin kx\right)\left(e^{-i\theta} \sin kx\right) = e^{-i\theta}e^{i\theta} \sin^2 kx = \sin^2 kx$ indistinguishable

17

c)

$$\cos(kx - \pi/2)\cos(kx - \pi/2) = \left(\left[\cos kx \cos \pi/2 + \sin kx \sin \pi/2\right]\right)^2$$
$$= \left(\left[\cos kx \times 0 + \sin kx \times 1\right]\right)^2 = \sin^2 kx \quad \text{indistinguishable}$$

indistinguishable

d)

$$i\cos(kx + \pi/2)(\sin\theta + i\cos\theta)\left(-\frac{\sqrt{2}}{2} + i\frac{\sqrt{2}}{2}\right)$$

$$\times(-i)\cos(kx + \pi/2)(\sin\theta - i\cos\theta)\left(-\frac{\sqrt{2}}{2} - i\frac{\sqrt{2}}{2}\right)$$

$$= \left[\cos(kx + \pi/2)\right]^2 (\sin^2\theta + \cos^2\theta)\left(\frac{1}{2} + \frac{1}{2}\right)$$

$$= \left[\cos(kx + \pi/2)\right]^2 = -\left[\cos(kx)\cos(\pi/2) + \sin(kx)\sin(\pi/2)\right]^2$$

$$= \sin^2(kx)$$

indistinguishable

P4.28) Is the superposition wave function $\psi(x) = \sqrt{2/a}\left[\sin(n\pi x/a) + \sin(m\pi x/a)\right]$ an eigenfunction of

the total energy operator for the particle in the box?

$$\hat{H}\psi(x) = -\frac{\hbar^2}{2m}\frac{d^2}{dx^2}\left[\sqrt{\frac{2}{a}}\sin\left(\frac{n\pi x}{a}\right) + \sqrt{\frac{2}{a}}\sin\left(\frac{m\pi x}{a}\right)\right]$$

$$= \frac{h^2 n^2}{8ma^2}\sqrt{\frac{2}{a}}\sin\left(\frac{n\pi x}{a}\right) + \frac{h^2 m^2}{8ma^2}\sqrt{\frac{2}{a}}\sin\left(\frac{m\pi x}{a}\right)$$

Because the result is not the wave function multiplied by a constant, the superposition wave function is

not an eigenfunction of the total energy operator.

P4.34) Calculate the probability that a particle in a one-dimensional box of length a is found between

$0.18a$ and $0.22a$ when it is described by the following wave functions:

a. $\sqrt{\dfrac{2}{a}}\sin\left(\dfrac{\pi x}{a}\right)$

b. $\sqrt{\dfrac{2}{a}}\sin\left(\dfrac{5\pi x}{a}\right)$

What would you expect for a classical particle? Compare your results in the two cases with the classical

result.

a)

Using the standard integral $\displaystyle\int \sin^2(by)\,dy = \dfrac{y}{2} - \dfrac{1}{4b}\sin(2by)$

$$P = \frac{2}{a}\int_{0.18a}^{0.22a} \sin^2\left(\frac{\pi x}{a}\right) dx = \frac{2}{a}\left[\frac{x}{2} - \frac{a}{4\pi}\sin\left(\frac{2\pi x}{a}\right)\right]_{0.18a}^{0.22a}$$

$$= \frac{2}{a}\left[\frac{0.22a}{2} - \frac{a}{4\pi}\sin(0.44\pi) - \frac{0.18a}{2} + \frac{a}{4\pi}\sin(0.36\pi)\right]$$

$$= 0.04 + \frac{1}{2\pi}\left[-\sin(0.44\pi) + \sin(0.36\pi)\right] = 0.028$$

b)

Using the standard integral $\displaystyle\int \sin^2(by)\,dy = \dfrac{y}{2} - \dfrac{1}{4b}\sin(2by)$

$$P = \frac{2}{a}\int_{0.18a}^{0.22a} \sin^2\left(\frac{5\pi x}{a}\right) dx = \frac{2}{a}\left[\frac{0.22a}{2} - \frac{a}{20\pi}\sin(2.2\pi) - \frac{0.31a}{2} + \frac{a}{20\pi}\sin(1.8\pi)\right]$$

$$= 0.04 + \frac{1}{10\pi}\left[-\sin(2.2\pi) + \sin(1.8\pi)\right] = 0.0026$$

Because a classical particle is equally likely to be in any given interval, the probability will be 0.04

independent of the energy. In the ground state, the interval chosen is near the maximum of the wave

function so that the quantum mechanical probability is greater than the classical probability. For the $n =$

3 state, the interval chosen is near a node of the wave function so that the quantum mechanical

probability is much less than the classical probability.

Chapter 5: The Particle in the Box and the Real World

Computational Problems

Before solving the computational problems, it is recommended that students work through Tutorials 1–3 under the Help menus in Spartan Student Edition to gain familiarity with the program.

Computational Problem 5.1: Build (a) ethylene, (b) the trans conformation for 1,3 butadiene, and (c) all trans hexatriene and calculate the ground-state (singlet) energy of these molecules using the B3LYP method with the 6-311+G** basis set. Repeat your calculation for the triplet state, which corresponds to the excitation of an electron from the highest filled energy level to the lowest unoccupied energy level. Use a nonplanar input geometry for the triplet states. Compare the energy difference from these calculations to literature values of the maximum in the UV-visible absorption spectrum.

Your solution to this problem should answer the questions in bold type in Steps 5 and 10.

Step 1: Create a new file, and build structures for 1,3-butadiene, 1,3,5-hexatriene, 1,3,5,7-octatetraene. Create each one as a new molecule within the same file using the "New Molecule" command in the File menu.

Step 2: Go to Setup > Calculations. Set the calculation type to "Equilibrium Geometry" and the method to B3LYP/6-31G*. The meaning of the method (Density Functional Theory w/B3LYP functional) and basis set (6-31G*) will be explained later in the course, consider this to be a "black box" for now. "Total Charge" should be neutral, and "Global Calculations" should be on. Click "OK." This will write the parameters for the calculation to the file, but will NOT start running the calculation. You will submit the calculation in Step 4.

Step 3: You will now calculate molecular orbitals for butadiene. Use the left and right navigation buttons at the bottom of the window to select your Butadiene model, then go to Setup → Surfaces. Click "Add," then select "LUMO" from the "Surfaces" menu. (NOT |LUMO| from the "Properties" menu!). Click "OK." Repeat for the HOMO, and HOMO-1. Make sure "Apply Globally" is NOT checked.

Step 4: Go to Setup → Submit. Your calculations should start running. The calculation will most likely take a few minutes on most current computers.

Step 5: While on butadiene, go to Display → Surfaces. A dialog box will appear. **Examine the surfaces one at a time, and comment on the degree of electron delocalization. Is there any evidence that the system can be modeled as particles in a 1-dimensional box?**

Step 6: Turn off the surfaces, close the dialog box, and then go to Display → Properties. Record the HOMO and LUMO energies in your spreadsheet program of choice. Use the navigation buttons on the bottom of the screen to advance to the next molecule, and repeat until you have recorded data for all 4 molecules.

The method used in this problem is reasonably accurate in calculating HOMO-LUMO gaps. Quantitatively accurate excited state calculations require advanced methods that are beyond the scope of this course. However, for this exercise, you will be looking for trends in the energy, not exact values, so use of a simpler method is acceptable.

Step 7: Calculate the energy gap E_g (in eV) by subtracting the HOMO energy from the LUMO energy. This is the energy corresponding to the longest-wavelength absorption of light by the conjugated molecule. Calculate the wavelength of absorption in another column using equation (1). Be careful about units here!

$$E = \frac{hc}{\lambda} \quad (1)$$

Step 8: Go through each molecule, measure the carbon-carbon bond lengths using the "Distance" tool in the tool palette, and add them up to obtain a total conjugation length for each molecule. Record this in your spreadsheet.

Step 9: Perform a 1-dimensional particle in-a-box calculation for the HOMO-LUMO transition energy for all three of your molecules, using the lengths you measured. Keep track of your units! Example values for 1, 3, 5, 7, 9-decapentaene have been provided to help get you started. The table below does not include wavelengths, however, you should calculate wavelengths for all four molecules. **If you are stuck here, please reread Section 5.3 of your book.**

Molecule	n HOMO	L (m)	B3LYP E_{HOMO} (eV)	B3LYP E_{LUMO} (eV)	B3LYP E_g (eV)	1D Box E_g (eV)
Butadiene						
Hexatriene						
Octatetraene						
Decapentaene	5	1.253×10^{-9}	-5.116	-1.796	3.32	2.64

Step 10: **Generate a table** containing the B3LYP and 1D Box predicted energy values, and **graph them** with # of double bonds on the x-axis and energy in eV on the y-axis. Look at the shape of the curves— **how does the first absorption energy change as a function of double bonds?** While you may use regression techniques if you would like, a qualitative answer is acceptable.

Also, compare your calculated values with the experimental value for butadiene and with some values that are calculated using a higher level theory. **What trends do you see?**

A UV/Visible absorption spectrum for butadiene is below. Find the strongest peak and use its energy in your comparison. Results for E_g obtained using a higher level theory for all four molecules are below.

B3LYP/6-31G* TDDFT results for

Butadiene: 5.27 eV

Hexatriene: 4.34 eV

Octatetraene: 3.72 eV

Decapentaene: 3.27 eV

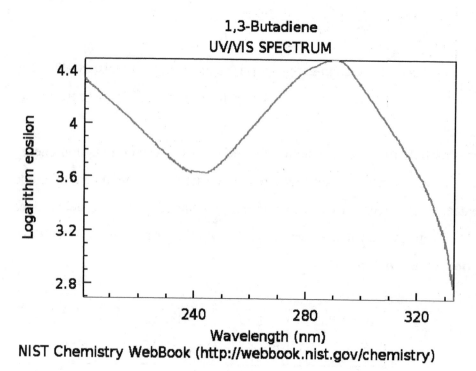

NIST Chemistry WebBook (http://webbook.nist.gov/chemistry)

Chapter 6: Commuting and Noncommuting Operators and the Surprising Consequences of Entanglement

P6.4) a) Show that $\psi(x) = e^{-x^2/2}$ is an eigenfunction of $\hat{A} = x^2 - \partial^2/\partial x^2$. Show that $\hat{B}\psi(x)$ where $\hat{B} = x - \partial/\partial x$ is another eigenfunction of $\hat{A}$.

$$\hat{A}\psi(x) = x^2 e^{-x^2/2} - \partial^2 e^{-x^2/2}/\partial x^2 = x^2 e^{-x^2/2} - \partial\left(-xe^{-x^2/2}\right)/\partial x$$

$$= e^{-x^2/2} + x^2 e^{-x^2/2} - x^2 e^{-x^2/2} = e^{-x^2/2}$$

$$\hat{B}\psi(x) = xe^{-x^2/2} - \partial e^{-x^2/2}/\partial x = xe^{-x^2/2} + xe^{-x^2/2} = 2xe^{-x^2/2}$$

$$\hat{A}\left(\hat{B}\psi(x)\right) = x^2\left(2xe^{-x^2/2}\right) - \partial^2\left(2xe^{-x^2/2}\right)/\partial x^2$$

$$= 2x^3 e^{-x^2/2} - \partial\left(2e^{-x^2/2} - 2x^2 e^{-x^2/2}\right)/\partial x$$

$$= 2x^3 e^{-x^2/2} + 2xe^{-x^2/2} + 4xe^{-x^2/2} - 2x^3 e^{-x^2/2}$$

$$= 6xe^{-x^2/2}$$

P6.5) Another important uncertainty principle is encountered in time-dependent systems. It relates the lifetime of a state Δt with the measured spread in the photon energy ΔE associated with the decay of this state to a stationary state of the system. "Derive" the relation $\Delta E\,\Delta t \geq \hbar/2$ in the following steps.

a. Starting from $E = p_x^2/2m$ and $\Delta E = (dE/dp_x)\Delta p_x$, show that $\Delta E = v_x\,\Delta p_x$.

b. Using $v_x = \Delta x/\Delta t$, show that $\Delta E\,\Delta t = \Delta p_x \Delta x \geq \hbar/2$.

c. Estimate the width of a spectral line originating from the decay of a state of lifetime 1.0×10^{-9} s and 1.0×10^{-11} s in inverse seconds and inverse centimeters.

a)

$$\frac{dE}{dp_x}\Delta p_x = \frac{p_x}{m}\Delta p_x = v_x\Delta p_x$$

b)

$$\Delta E\Delta t = \frac{\Delta x}{\Delta t}\Delta p_x\Delta t = \Delta x\Delta p_x \geq \frac{\hbar}{2}$$

c)

$$\Delta E \geq \frac{\hbar}{2\Delta t} = h\Delta \nu$$

$$\Delta \nu = \frac{1}{4\pi\Delta t} = \frac{1}{4\pi\left(1.0\times10^{-9}s\right)} = 8.0\times10^{7} \ s^{-1}$$

$$\Delta \nu \left(cm^{-1}\right) = \frac{\Delta \nu \left(s^{-1}\right)}{c} = \frac{8.0\times10^{7} \ s^{-1}}{2.998\times10^{10} \ cm \ s^{-1}} = 0.00265 \ cm^{-1}$$

The corresponding answers for 1.0×10^{-11} s are 8.0×10^{9} s^{-1} and 0.265 cm^{-1}, respectively.

P6.6) Evaluate the commutator $[x(\partial/\partial y), y]$ by applying the operators to an arbitrary function $f(x,y)$.

$$\left[x\frac{\partial}{\partial y}, y\right]f(x,y) = x\frac{\partial yf(x,y)}{\partial y} - yx\frac{\partial f(x,y)}{\partial y}$$

$$= xf(x,y) + xy\frac{\partial f(x,y)}{\partial y} - yx\frac{\partial f(x,y)}{\partial y} = xf(x,y)$$

Therefore $\left[x\dfrac{\partial}{\partial y}, y\right] = x$

P6.8) Consider the entangled wave function for two photons,

$$\psi_{12} = \frac{1}{\sqrt{2}}(\psi_1(H)\psi_2(V) + \psi_1(V)\psi_2(H))$$

Assume that the polarization operator $\hat{P}_i$ has the properties $\hat{P}_i\psi_i(H) = -\psi_i(H)$ and $\hat{P}_i\psi_i(V) = +\psi_i(V)$ where $i = 1$ or $i = 2$.

a. Show that ψ_{12} is not an eigenfunction of $\hat{P}_1$ or $\hat{P}_2$.

b. Show that each of the two terms in ψ_{12} is an eigenfunction of the polarization operator $\hat{P}_1$.

c. What is the average value of the polarization P_1 that you will measure on identically prepared systems? It is not necessary to do a calculation to answer this question.

a)

$$\hat{P}_1 \psi_{12} = \frac{1}{\sqrt{2}} (-\psi_1(H)\psi_2(V) + \psi_1(V)\psi_2(H))$$

$$\hat{P}_2 \psi_{12} = \frac{1}{\sqrt{2}} (\psi_1(H)\psi_2(V) - \psi_1(V)\psi_2(H))$$

In neither case does the operation return the original function multiplied by a constant. Therefore, the function is not an eigenfunction of either operator.

b) Each of the 2 terms of the above expression is the original term multiplied by a constant. Therefore each individual term is an eigenfunction of the operators.

c) A measurement will project the system into the wave function $\psi_1(H)\psi_2(V)$ or $-\psi_1(V)\psi_2(H)$ with equal probability. Therefore it is equally likely to measure the eigenvalue +1 as −1 and the average of the measured values will be zero.

P6.10) Revisit the double-slit experiment of Example Problem 6.2. Using the same geometry and relative uncertainty in the momentum, what electron momentum would give a position uncertainty of 1.00×10^{-9} m? What is the ratio of the wavelength and the slit spacing for this momentum? Would you expect a pronounced diffraction effect for this wavelength?

$$\Delta p = \frac{1}{2} \frac{\hbar}{\Delta x} = \frac{1.055 \times 10^{-34} \text{ J s}}{2.00 \times 10^{-9} \text{ m}} = 5.275 \times 10^{-26} \text{ kg m s}^{-1}$$

$$p = \frac{\Delta p}{0.01} = 5.275 \times 10^{-24} \text{ kg m s}^{-1}$$

$$\lambda = \frac{h}{p} = \frac{6.626 \times 10^{-34} \text{ J s}}{5.275 \times 10^{-24} \text{ kg m s}^{-1}} = 1.26 \times 10^{-10} \text{ m}$$

Because $\dfrac{\lambda}{b} = \dfrac{1.26 \times 10^{-10} \text{ m}}{1.000 \times 10^{-9} \text{ m}} = 0.126$, where b is the slit spacing, the diffraction will hardly be noticeable.

P6.16) Evaluate the commutator $[d/dx, x^2]$ by applying the operators to an arbitrary function $f(x)$.

$$\left[\frac{d}{dx}, x^2\right] f(x) = \frac{d}{dx}\left(x^2 f(x)\right) - x^2 \frac{d}{dx} f(x)$$

$$= 2xf(x) + x^2 \frac{d}{dx} f(x) - x^2 \frac{d}{dx} f(x) = 2xf(x)$$

$$\left[\frac{d}{dx}, x^2\right] = 2x$$

P6.19) Evaluate the commutator $\left[(d/dx) - x, (d/dx) + x\right]$ by applying the operators to an arbitrary function $f(x)$.

$$\left[\frac{d}{dx} - x, \frac{d}{dx} + x\right] f(x) = \left(\frac{d}{dx} - x\right)\left(\frac{d}{dx} + x\right) f(x) - \left(\frac{d}{dx} + x\right)\left(\frac{d}{dx} - x\right) f(x)$$

$$= \left(\frac{d}{dx} - x\right)\left(\frac{df(x)}{dx} + xf(x)\right) - \left(\frac{d}{dx} + x\right)\left(\frac{df(x)}{dx} - xf(x)\right)$$

$$= \frac{d^2 f(x)}{dx^2} + f(x) + x\frac{df(x)}{dx} - x\frac{df(x)}{dx} - x^2 f(x) - \frac{d^2 f(x)}{dx^2} + f(x) + x\frac{df(x)}{dx}$$

$$-x\frac{df(x)}{dx} + x^2 f(x)$$

$$= 2f(x) \text{ Therefore,}$$

$$\left[\frac{d}{dx} - x, \frac{d}{dx} + x\right] = 2$$

P6.23) The muzzle velocity of a rifle bullet is about 775 m s^{-1}. If the bullet weighs 28 g, and the uncertainty in its momentum is 0.15%, how accurately can the position of the bullet be measured?

$$p = mv = 28 \times 10^{-3} \text{ kg} \times 775 \text{ m s}^{-1} = 22 \text{ kg m s}^{-1}$$

$$\Delta p = 10^{-3} p = 2.2 \times 10^{-2} \text{ kg m s}^{-1}$$

$$\Delta x = \frac{\hbar}{2\Delta p} = \frac{1.055 \times 10^{-34} \text{ J s}}{2 \times 2.2 \times 10^{-2} \text{ kg m s}^{-1}} = 1.6 \times 10^{-33} \text{ m}$$

Chapter 7: A Quantum Mechanical Model for the Vibration and Rotation of Molecules

P7.1) A gas-phase $^1H^{35}Cl$ molecule, with a bond length of 127.5 pm, rotates in a three-dimensional space.

a. Calculate the zero point energy associated with this rotation.

b. What is the smallest quantum of energy that can be absorbed by this molecule in a rotational excitation?

a) There is no zero point energy because the rotation is not constrained.

b) The smallest energy that can be absorbed is

$$E = \frac{\hbar^2}{2I}J(J+1) = \frac{\hbar^2}{2I}1(1+1)$$

$$= \frac{2\times\left(1.055\times10^{-34}\text{ J s}\right)^2}{2\times\dfrac{1.0078\text{ amu}\times34.9688\text{ amu}}{1.0078\text{ amu}+34.9688\text{ amu}}\times1.66\times10^{-27}\text{ kg amu}^{-1}\times\left(127.5\times10^{-12}\text{ m}\right)^2}$$

$$E = 4.21\times10^{-22}\text{ J}$$

P7.3) Using the Boltzmann distribution, calculate n_J/n_0 for $^1H^{19}F$ for $J = 0, 5, 10,$ and 20 at $T = 650$. K. Does n_J/n_0 go through a maximum as J increases? If so, what can you say about the value of J corresponding to the maximum?

$$\frac{n_J}{n_0} = (2J+1)e^{-(E_J-E_0)/kT} = (2J+1)\exp[-E_J/kT]$$

$$I = \mu r_0^2 = \frac{1.0078 \text{ amu} \times 18.9984 \text{ amu}}{1.0078 \text{ amu} + 18.994 \text{ amu}} \times 1.66 \times 10^{-27} \text{ kg amu}^{-1} \times (91.68 \times 10^{-12} \text{ m})^2$$

$$= 1.34 \times 10^{-47} \text{ kg m}^2$$

$$E = \frac{\hbar^2}{2I}J(J+1) = \frac{(1.055 \times 10^{-34} \text{ J s})^2}{2 \times 1.34 \times 10^{-47} \text{ kg m}^2}J(J+1) = 4.17 \times 10^{-22} J(J+1)$$

$$\frac{n_0}{n_0} = 1$$

$$\frac{n_5}{n_0} = (2 \times 5 + 1)\exp\left[-(30 \times 4.17 \times 10^{-22} \text{ J})/1.381 \times 10^{-23} \text{ J K}^{-1} \times 650. \text{ K}\right] = 2.73$$

$$\frac{n_{10}}{n_0} = (2 \times 10 + 1)\exp\left[-(110 \times 4.17 \times 10^{-22} \text{ J})/1.381 \times 10^{-23} \text{ J K}^{-1} \times 650. \text{ K}\right] = 0.127$$

$$\frac{n_{20}}{n_0} = (2 \times 20 + 1)\exp\left[-(420 \times 4.17 \times 10^{-22} \text{ J})/1.381 \times 10^{-23} \text{ J K}^{-1} \times 650. \text{ K}\right] = 1.40 \times 10^{-7}$$

$\dfrac{n_J}{n_0}$ goes through a maximum because it has a value greater than one for $J = 5$. You can only conclude that $J_{max} \leq 10$.

P7.8) The vibrational frequency for $^{19}F_2$ expressed in wave numbers is 916.64 cm^{-1}. What is the force constant associated with the F-F bond? How much would a classical spring with this force constant be elongated if a mass of 2.50 kg were attached to it? Use the gravitational acceleration on Earth at sea level for this problem.

$$v = c\tilde{v} = \frac{1}{2\pi}\sqrt{\frac{k}{\mu}}$$

so $k = (2\pi c\tilde{v})^2 \mu$

$$k = (2\pi \times 2.998 \times 10^{10} \text{ cm s}^{-1} \times 916.64 \text{ cm}^{-1})^2 \times \frac{18.94 \text{ amu} \times 18.94 \text{ amu}}{2 \times 18.94 \text{ amu}} \times \frac{1.661 \times 10^{-27} \text{ kg}}{\text{amu}}$$

$$k = 470. \text{ N m}^{-1}$$

$$x = \frac{F}{k} = \frac{mg}{k} = \frac{2.50 \text{ kg} \times 9.81 \text{ m s}^{-2}}{470. \text{ N m}^{-1}} = 5.21 \times 10^{-2} \text{ m}$$

P7.9) Calculate E_{rot}/kT for $H^{19}F$ for $J = 0, 5, 10,$ and 20 at 298 K. For which of these values of J is $E_{rot}/kT \geq 10$?

$$I = \mu r_0^2 = \frac{1.0078 \times 18.9984}{1.0078 + 18.994} \times 1.66 \times 10^{-27} \text{ kg amu}^{-1} \times \left(91.68 \times 10^{-12} \text{ m}\right)^2 = 1.34 \times 10^{-47} \text{ kg m}^2$$

$$E = \frac{\hbar^2}{2I} J(J+1) = \frac{\left(1.055 \times 10^{-34} \text{ J s}\right)^2}{2 \times 1.34 \times 10^{-47} \text{ kg m}^2} J(J+1) = 4.17 \times 10^{-22} J(J+1)$$

$$E_{J=0} = 0$$

$$E_{J=5} = 30 \times 4.17 \times 10^{-22} \text{ J} = 12.5 \times 10^{-21} \text{ J}$$

$$\frac{E_{J=5}}{kT} = \frac{12.5 \times 10^{-21} \text{ J}}{1.381 \times 10^{-23} \text{ J K}^{-1} \times 298 \text{ K}} = 3.04$$

$$E_{J=10} = 110 \times 4.17 \times 10^{-22} \text{ J} = 4.58 \times 10^{-20} \text{ J}$$

$$\frac{E_{J=10}}{kT} = \frac{4.58 \times 10^{-20} \text{ J}}{1.381 \times 10^{-23} \text{ J K}^{-1} \times 298 \text{ K}} = 11.1$$

$$E_{J=20} = 20 \times 21 \times 4.17 \times 10^{-22} \text{ J} = 1.75 \times 10^{-19} \text{ J}$$

$$\frac{E_{J=20}}{kT} = \frac{17.5 \times 10^{-20} \text{ J}}{1.381 \times 10^{-23} \text{ J K}^{-1} \times 298 \text{ K}} = 42.5$$

$$\frac{E_{rot}}{kT} > 10 \text{ for } J = 10 \text{ and } 20.$$

P7.12) Show by carrying out the appropriate integration that the total energy eigenfunctions for the harmonic oscillator $\psi_0(x) = (\alpha/\pi)^{1/4} e^{-(1/2)\alpha x^2}$ and $\psi_2(x) = (\alpha/4\pi)^{1/4}(2\alpha x^2 - 1)e^{-(1/2)\alpha x^2}$ are orthogonal over the interval $-\infty < x < \infty$ and that $\psi_2(x)$ is normalized over the same interval. In evaluating integrals of this type, $\int_{-\infty}^{\infty} f(x)\,dx = 0$ if $f(x)$ is an odd function of x and $\int_{-\infty}^{\infty} f(x)\,dx = 2\int_0^{\infty} f(x)\,dx$ if $f(x)$ is an even function of x.

We use the standard integrals $\int_0^{\infty} x^{2n} e^{-ax^2}\,dx = \frac{1 \cdot 3 \cdot 5 \cdots (2n-1)}{2^{n+1} a^n} \sqrt{\frac{\pi}{a}}$ and

$$\int_0^{\infty} e^{-ax^2}\,dx = \left(\frac{\pi}{4a}\right)^{1/2}$$

$$\int_{-\infty}^{\infty} \psi_2^*(x)\psi_0(x)dx = \int_{-\infty}^{\infty} \left(\frac{\alpha}{4\pi}\right)^{1/4}\left(2\alpha x^2 - 1\right)e^{-\frac{1}{2}\alpha x^2}\left(\frac{\alpha}{\pi}\right)^{1/4}e^{-\frac{1}{2}\alpha x^2}dx$$

$$= \left(\frac{\alpha^2}{4\pi^2}\right)^{1/4}\int_{-\infty}^{\infty}\left(2\alpha x^2 - 1\right)e^{-\alpha x^2}dx = 2\left(\frac{\alpha^2}{4\pi^2}\right)^{1/4}\int_{0}^{\infty}\left(2\alpha x^2 - 1\right)e^{-\alpha x^2}dx$$

$$= \left(\frac{\alpha^2}{4\pi^2}\right)^{1/4}\left(2\alpha\frac{1}{4\alpha}\sqrt{\frac{\pi}{\alpha}} - \frac{1}{2}\sqrt{\frac{\pi}{\alpha}}\right) = 0$$

$$\int_{-\infty}^{\infty} \psi_2^*(x)\psi_2(x)dx = \int_{-\infty}^{\infty} \left(\frac{\alpha}{4\pi}\right)^{1/4}\left(2\alpha x^2 - 1\right)e^{-\frac{1}{2}\alpha x^2}\left(\frac{\alpha}{4\pi}\right)^{1/4}\left(2\alpha x^2 - 1\right)e^{-\frac{1}{2}\alpha x^2}dx$$

$$= 2\left(\frac{\alpha}{4\pi}\right)^{1/2}\int_{0}^{\infty}\left(4\alpha^2 x^4 - 4\alpha x^2 + 1\right)e^{-\alpha x^2}dx$$

$$= 2\left(\frac{\alpha}{4\pi}\right)^{1/2}\left(4\alpha^2\frac{3}{2^3\alpha^2}\sqrt{\frac{\pi}{\alpha}} - 4\alpha\frac{1}{2^2\alpha}\sqrt{\frac{\pi}{\alpha}} + \frac{1}{2}\sqrt{\frac{\pi}{\alpha}}\right) = 2\left(\frac{\alpha}{4\pi}\right)^{1/2}\sqrt{\frac{\pi}{\alpha}}\left(\frac{3}{2} - 1 + \frac{1}{2}\right) = 1$$

P7.14) Calculate the frequency and wavelength of the radiation absorbed when a quantum harmonic oscillator with a frequency of 5.58×10^{13} s^{-1} makes a transition from the $n = 3$ to the $n = 4$ state.

$$\Delta E = \left(4 + 1/2\right)h\nu - \left(3 + 1/2\right)h\nu = h\nu$$
$$= 6.626 \times 10^{-34} \text{ J s} \times 5.58 \times 10^{13} \text{ s}^{-1} = 3.70 \times 10^{-20} \text{ J}$$
$$\nu = \frac{E}{h} = 5.58 \times 10^{13} \text{ s}^{-1}$$
$$\lambda = \frac{c}{\nu} = \frac{2.998 \times 10^8 \text{ m s}^{-1}}{5.58 \times 10^{13} \text{ s}^{-1}} = 5.37 \times 10^{-6} \text{ m}$$

P7.16) The vibrational frequency of ^{1}H^{19}F is 1.24×10^{14} s^{-1}. Calculate the force constant of the molecule. How large a mass would be required to stretch a classical spring with this force constant by 1.00 cm? Use the gravitational acceleration on Earth at sea level for this problem.

$$\nu = \frac{1}{2\pi}\sqrt{\frac{k}{\mu}}; \quad k = 4\pi^2 \mu \nu^2$$

$$k = 4 \times \pi^2 \times \frac{1.008 \text{ amu} \times 18.9984 \text{ amu}}{1.008 \text{ amu} + 18.9984 \text{ amu}} \times \frac{1.661 \times 10^{-27} \text{ kg}}{\text{amu}} \times \left(1.24 \times 10^{14} \text{ s}^{-1}\right)^2$$

$$k = 965 \text{ kg s}^{-2}$$

$$F = kx = mg$$

$$m = \frac{kx}{g} = \frac{965 \text{ kg s}^{-2} \times 10^{-2} \text{ m}}{9.81 \text{ m s}^{-2}} = 0.984 \text{ kg}$$

P7.17) Use $\sqrt{\langle x^2 \rangle}$ as calculated in Problem P7.32 as a measure of the vibrational amplitude for a molecule. What fraction is $\sqrt{\langle x^2 \rangle}$ of the 141.4-pm bond length of the $^1H^{81}Br$ molecule for $n = 0, 1,$ and 2? The force constant for the $^1H^{81}Br$ molecule is 412 N m^{-1}.

$$\sqrt{\langle x^2 \rangle} = \left(\frac{\hbar}{2\sqrt{k\mu}}\right)^{\frac{1}{2}} = \left(\frac{1.055 \times 10^{-34} \text{ J s}}{2 \times \sqrt{412 \text{ N m}^{-1} \times \frac{1.0078 \times 80.9163}{1.0078 + 80.9163} \times 1.66 \times 10^{-27} \text{ kg amu}^{-1}}}\right)^{\frac{1}{2}}$$

For $n = 0$, $= 8.00 \times 10^{-12}$ m

$$\frac{\sqrt{\langle x^2 \rangle}}{\text{bond length}} = \frac{8.00 \times 10^{-12} \text{ m}}{141.4 \times 10^{-12} \text{ m}} = 0.0565$$

For $n = 1$

$$\sqrt{\langle x^2 \rangle} = \left(\frac{3\hbar}{2\sqrt{k\mu}}\right)^{\frac{1}{2}}$$

$$= \left(\frac{3 \times 1.055 \times 10^{-34} \text{ J s}}{2 \times \sqrt{412 \text{ N m}^{-1} \times \frac{1.0078 \text{ amu} \times 80.9163 \text{ amu}}{1.0078 \text{ amu} + 80.9163 \text{ amu}} \times 1.66 \times 10^{-27} \text{ kg amu}^{-1}}}\right)^{\frac{1}{2}}$$

$$= 1.38 \times 10^{-11} \text{ m}$$

$$\frac{\sqrt{\langle x^2 \rangle}}{\text{bond length}} = \frac{1.38 \times 10^{-11} \text{ m}}{141.4 \times 10^{-12} \text{ m}} = 0.0979$$

For $n = 2$

$$\sqrt{\langle x^2 \rangle} = \left(\frac{5\hbar}{2\sqrt{k\mu}} \right)^{\frac{1}{2}}$$

$$= \left(\frac{5 \times 1.055 \times 10^{-34} \text{ J s}}{2 \times \sqrt{412 \text{ N m}^{-1} \times \frac{1.0078 \text{ amu} \times 80.9163 \text{ amu}}{1.0078 \text{ amu} + 80.9163 \text{ amu}} \times 1.66 \times 10^{-27} \text{ kg amu}^{-1}}} \right)^{\frac{1}{2}}$$

$$= 1.79 \times 10^{-11} \text{ m}$$

$$\frac{\sqrt{\langle x^2 \rangle}}{\text{bond length}} = \frac{1.79 \times 10^{-11} \text{ m}}{141.4 \times 10^{-12} \text{ m}} = 0.126$$

P7.18) A coin with a mass of 5.67 g suspended on a rubber band has a vibrational frequency of 3.00 s^{-1}. Calculate a) the force constant of the rubber band; b) the zero point energy; c) the total vibrational energy if the maximum displacement is 0.500 cm; and d) the vibrational quantum number corresponding to the energy in part c).

$$v = \frac{1}{2\pi}\sqrt{\frac{k}{m}} \quad k = 4\pi^2 v^2 \mu$$

$$k = 4\pi^2 \times 9.00 \text{ s}^{-2} \times 5.67 \times 10^{-3} \text{ kg} = 2.01 \text{ N m}^{-1}$$

$$E_0 = \frac{1}{2}hv = 6.626 \times 10^{-34} \text{ J s} \times 3.00 \text{ s}^{-1} = 9.94 \times 10^{-34} \text{ J}$$

$$E_n = \frac{1}{2}kx_{max}^2 = \frac{1}{2} \times 2.01 \text{ N m}^{-1} \times \left(0.500 \times 10^{-2} \text{ m}\right)^2 = 2.52 \times 10^{-5} \text{ J}$$

$$E_n = \left(n + \frac{1}{2}\right)hv \quad n = \frac{E_n}{hv} - \frac{1}{2}$$

$$n = \frac{2.52 \times 10^{-5} \text{ J}}{6.626 \times 10^{-34} \text{ J s} \times 3.00 \text{ s}^{-1}} - \frac{1}{2} = 1.27 \times 10^{28}$$

P7.21) Is it possible to simultaneously know the angular orientation of a molecule rotating in a two-dimensional space and its angular momentum? Answer this question by evaluating the commutator $[\phi, -i\hbar(\partial/\partial\phi)]$.

$$\left[\phi, -i\hbar\frac{\partial}{\partial\phi}\right]f(\phi) = -i\hbar\phi\frac{df(\phi)}{d\phi} + i\hbar\frac{d\left[\phi f(\phi)\right]}{d\phi}\phi = i\hbar f(\phi)$$

$$\left[\phi, -i\hbar\frac{\partial}{\partial\phi}\right] = i\hbar$$

Because the commutator is not equal to zero, it is not possible to simultaneously know the angular orientation of a molecule rotating in a two-dimensional space and its angular momentum.

P7.23) The force constant for a $H^{35}Cl$ molecule is 516 N m^{-1}.

a. Calculate the zero point vibrational energy for this molecule for a harmonic potential.

b. Calculate the light frequency needed to excite this molecule from the ground state to the first excited state.

a)

$$E_1 = h\sqrt{\frac{k}{\mu}}\left(1+\frac{1}{2}\right) = \frac{3}{2}\times1.055\times10^{-34}\text{ J s}\times\sqrt{\frac{516\text{ N m}^{-1}}{\frac{1.0078\text{ amu}\times34.9688\text{ amu}}{1.0078\text{ amu} + 34.9688\text{ amu}}\times1.66\times10^{-27}\text{ kg amu}^{-1}}}$$

$$E_1 = 8.91\times10^{-20}\text{ J}$$

$$E_0 = \hbar\sqrt{\frac{k}{\mu}}\left(\frac{1}{2}\right) = \frac{1}{3}E_1 = 2.97\times10^{-20}\text{ J}$$

b) $v = \dfrac{E_1 - E_0}{h} = \dfrac{8.91\times10^{-20}\text{ J} - 2.97\times10^{-20}\text{ J}}{6.626\times10^{-34}\text{ J s}} = 8.97\times10^{13}\text{ s}^{-1}$

P7.25) A $^1H^{127}I$ molecule, with a bond length of 160.92 pm, absorbed on a surface rotates in two dimensions.

a. Calculate the zero point energy associated with this rotation.

b. What is the smallest quantum of energy that can be absorbed by this molecule in a rotational excitation?

a) There is no zero point energy because the rotation is not constrained.

b) The smallest energy that can be absorbed is

$$E = \frac{\hbar^2 m_l^2}{2I} = \frac{\hbar^2}{2I} = \frac{\left(1.055\times10^{-34}\text{ J s}\right)^2}{2\times\frac{1.0078\text{ amu}\times126.9045\text{ amu}}{1.0078\text{ amu} + 126.9045\text{ amu}}\times1.66\times10^{-27}\text{ kg amu}^{-1}\times\left(160.92\times10^{-12}\text{ m}\right)^2}$$

$$E = 1.29\times10^{-22}\text{ J}$$

P7.29) Evaluate the average linear momentum of the quantum harmonic oscillator, $\langle p_x \rangle$, for the ground state ($n = 0$) and first two excited states ($n = 1$ and $n = 2$). Use the hint about evaluating integrals in Problem P7.12.

We use the standard integrals $\int_0^\infty x^{2n} e^{-ax^2} dx = \dfrac{1 \cdot 3 \cdot 5 \ldots (2n-1)}{2^{n+1} a^n} \sqrt{\dfrac{\pi}{a}}$ and

$$\int_0^\infty e^{-ax^2} dx = \left(\frac{\pi}{4a} \right)^{1/2}$$

$$\langle p_x \rangle = \int_{-\infty}^\infty \psi_n^*(x) \left(-i\hbar \frac{d}{dx} \right) \psi_n \, dx$$

for $n = 0$, $\langle p_x \rangle = \int_{-\infty}^\infty \left(\frac{\alpha}{\pi} \right)^{1/4} e^{-\frac{1}{2}\alpha x^2} \left(-i\hbar \frac{d}{dx} \right) \left(\frac{\alpha}{\pi} \right)^{1/4} e^{-\frac{1}{2}\alpha x^2} \, dx$

$$\langle p_x \rangle = \left(\frac{\alpha}{\pi} \right)^{1/2} (-i\hbar) \int_{-\infty}^\infty -\alpha x e^{-\alpha x^2} \, dx$$

Because the integrand is an odd function of x, $\langle p_x \rangle = 0$ for $n = 0$.

for $n = 1$, $\langle p_x \rangle = \int_{-\infty}^\infty \left(\frac{4\alpha^3}{\pi} \right)^{1/4} x e^{-\frac{1}{2}\alpha x^2} \left(-i\hbar \frac{d}{dx} \right) \left(\frac{4\alpha^3}{\pi} \right)^{1/4} x e^{-\frac{1}{2}\alpha x^2} \, dx$

$$\langle p_x \rangle = \left(\frac{4\alpha^3}{\pi} \right)^{1/2} (-i\hbar) \int_{-\infty}^\infty x \left(1 - \alpha x^2 \right) e^{-\alpha x^2} \, dx$$

Because the integrand is an odd function of x, $\langle p_x \rangle = 0$ for $n = 1$.

for $n = 2$, $\langle p_x \rangle = \int_{-\infty}^\infty \left(\frac{\alpha}{4\pi} \right)^{1/4} \left(2\alpha x^2 - 1 \right) e^{-\frac{1}{2}\alpha x^2} \left(-i\hbar \frac{d}{dx} \right) \left(\frac{\alpha}{4\pi} \right)^{1/4} \left(2\alpha x^2 - 1 \right) e^{-\frac{1}{2}\alpha x^2} \, dx$

$$\langle p_x \rangle = \left(\frac{\alpha}{4\pi} \right)^{1/2} (-i\hbar) \int_{-\infty}^\infty \left(2\alpha x^2 - 1 \right) e^{-\alpha x^2} \left(-2\alpha^2 x^3 + 4\alpha x + \alpha x \right) dx$$

$$\langle p_x \rangle = \left(\frac{\alpha}{4\pi} \right)^{1/2} (-i\hbar) \int_{-\infty}^\infty e^{-\alpha x^2} \left(-4\alpha^3 x^5 + 12\alpha^2 x^3 - 5\alpha x \right) dx$$

Because the integrand is an odd function of x, $\langle p_x \rangle = 0$ for $n = 3$.

The result is general. $\langle p_x \rangle = 0$ for all values of n.

P7.33) For molecular rotation, the symbol J rather than l is used as the quantum number for angular momentum. A $^1\text{H}^{19}\text{F}$ molecule has the rotational quantum number $J = 10$ and vibrational quantum number $n = 0$.

a. Calculate the rotational and vibrational energy of the molecule. Compare each of these energies with kT at 300. K.

b. Calculate the period for vibration and rotation. How many times does the molecule rotate during one vibrational period?

a)

$$E_{rot} = \frac{J(J+1)\hbar^2}{2\mu r^2} = \frac{10 \times 11 \times \left(1.055 \times 10^{-34} \text{ J s}\right)^2}{2 \times \dfrac{1.0078 \text{ amu} \times 18.994 \text{ amu}}{1.0078 \text{ amu} + 18.994 \text{ amu}} \times \dfrac{1.661 \times 10^{-27} \text{ kg}}{\text{amu}} \times \left(91.68 \times 10^{-12} \text{ m}\right)^2}$$

$$E_{rot} = 4.58 \times 10^{-20} \text{ J}$$

$$\frac{E_{rot}}{kT} = \frac{4.58 \times 10^{-20} \text{ J}}{1.381 \times 10^{-23} \text{ J K}^{-1} \times 300. \text{ K}} = 11.1$$

$$E_{vib} = \left(n + \frac{1}{2}\right)\hbar\sqrt{\frac{k}{\mu}} = \frac{1}{2} \times 1.055 \times 10^{-34} \text{ J s} \times \sqrt{\frac{966 \text{ N m}^{-1}}{\dfrac{1.0078 \text{ amu} \times 18.994 \text{ amu}}{1.0078 \text{ amu} + 18.994 \text{ amu}} \times \dfrac{1.661 \times 10^{-27} \text{ kg}}{\text{amu}}}}$$

$$E_{vib} = 4.11 \times 10^{-20} \text{ J}$$

$$\frac{E_{vib}}{kT} = \frac{4.11 \times 10^{-20} \text{ J}}{1.381 \times 10^{-23} \text{ J K}^{-1} \times 300. \text{ K}} = 9.93$$

b) Calculate the period for vibration and rotation. How many times does the molecule vibrate during one rotational period?

$$E_{rot} = \frac{1}{2}I\omega^2; \quad \omega = 2\pi v = \frac{2\pi}{T_{rot}}$$

$$T_{rot} = \frac{2\pi}{\omega} = \frac{2\pi}{\sqrt{\dfrac{2E_{rot}}{I}}}$$

$$= \frac{2\pi}{\sqrt{\dfrac{2 \times 4.58 \times 10^{-20} \text{ J}}{\dfrac{1.0078 \text{ amu} \times 18.994 \text{ amu}}{1.0078 \text{ amu} + 18.994 \text{ amu}} \times \dfrac{1.661 \times 10^{-27} \text{ kg}}{\text{amu}} \times \left(91.68 \times 10^{-12} \text{ m}\right)^2}}} = 7.59 \times 10^{-14} \text{ s}$$

$$T_{vib} = \frac{1}{v} = 2\pi\sqrt{\frac{\mu}{k}} = 2\pi\sqrt{\frac{\dfrac{1.0078 \text{ amu} \times 18.994 \text{ amu}}{1.0078 \text{ amu} + 18.994 \text{ amu}} \times \dfrac{1.661 \times 10^{-27} \text{ kg}}{\text{amu}}}{966 \text{ N m}^{-1}}} = 8.06 \times 10^{-15} \text{ s}$$

It vibrates $\dfrac{T_{rot}}{T_{vib}} = \dfrac{7.59 \times 10^{-14} \text{ s}}{8.06 \times 10^{-15} \text{ s}} = 9.41$ times in one rotational period.

Chapter 8: The Vibrational and Rotational Spectroscopy of Diatomic Molecules

P8.2) The infrared spectrum of $^7Li^{35}Cl$ has an intense line at 643 cm^{-1}. Calculate the force constant and period of vibration of this molecule.

$$k = 4\pi^2 v^2 \mu$$

$$= 4\pi^2 \left(2.998 \times 10^{10} \text{ cm s}^{-1} \times 643 \text{ cm}^{-1}\right)^2 \times \frac{34.9688 \text{ amu} \times 7.0160 \text{ amu}}{34.9688 \text{ amu} + 7.0160 \text{ amu}} \times 1.6605 \times 10^{-27} \text{ kg amu}^{-1}$$

$$= 142 \text{ N m}^{-1}$$

$$T = \frac{1}{v} = 5.19 \times 10^{-14} \text{ s}$$

P8.3) Purification of water for drinking using UV light is a viable way to provide potable water in many areas of the world. Experimentally, the decrease in UV light of wavelength 250 nm follows the empirical relation $I/I_0 = e^{-\varepsilon' l}$ where l is the distance that the light passed through the water and ε' is an effective absorption coefficient. $\varepsilon' = 0.070$ cm^{-1} for pure water and 0.30 cm^{-1} for water exiting a waste water treatment plant. What distance corresponds to a decrease in I of 10.% from its incident value for a) pure water and b) waste water?

$$\ln\left[\frac{I(\lambda)}{I_0(\lambda)}\right] = -\varepsilon'(\lambda)l; \quad l = -\frac{\ln\left[\frac{I(\lambda)}{I_0(\lambda)}\right]}{\varepsilon'}$$

$$l = \frac{\ln[0.90]}{0.070 \text{ cm}^{-1}} = 1.5 \text{ cm for pure water}$$

$$l = \frac{\ln[0.90]}{0.30 \text{ cm}^{-1}} = 0.35 \text{ cm for treatment plant water}$$

P8.8) An infrared absorption spectrum of an organic compound is shown in the following figure. Use the characteristic group frequencies listed in Section 8.5 to decide whether this compound is more likely to be ethyl amine, pentanol, or acetone.

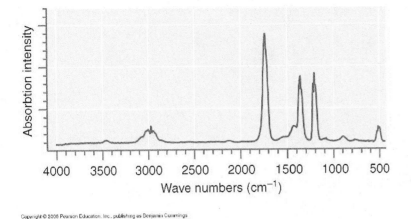

The major peak near 1700 cm^{-1} is the C=O stretch and the peak near 1200 cm^{-1} is a C–C–C stretch. These peaks are consistent with the compound being acetone. Ethyl amine should show a strong peak near 3350 cm^{-1} and pentanol should show a strong peak near 3400 cm^{-1}. Because these peaks are absent, these compounds can be ruled out.

P8.10) Write an expression for the moment of inertia of the acetylene molecule in terms of the bond distances. Does this molecule have a pure rotational spectrum?

The two mutually perpendicular axes of rotation are perpendicular to the molecular axis and go through the center of the molecule. The two moments of inertia are equal.

$$ I = \sum_i m_i x_i^2 = 2\left(m_C \left[\frac{x_{C\equiv C}}{2} \right]^2 + m_H \left[\frac{x_{C\equiv C}}{2} + x_{C-H} \right]^2 \right) $$

Because the molecule has no dipole moment, it does not have a pure rotational spectrum.

P8.15) Calculating the motion of individual atoms in the vibrational modes of molecules (called normal modes) is an advanced topic. Given the normal modes shown in the following figure, decide which of the normal modes of CO_2 and H_2O have a nonzero dynamical dipole moment and are therefore infrared active. The motion of the atoms in the second of the two doubly degenerate bend modes for CO_2 is identical to the first, but is perpendicular to the plane of the page.

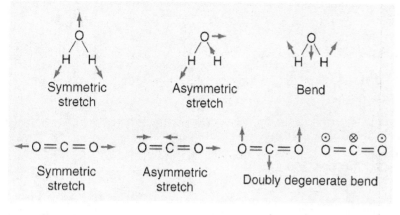

All three vibrational modes of water will lead to a change in the dipole moment and are therefore infrared active. The symmetric stretch of carbon dioxide will not lead to a change in the dipole moment and is infrared inactive. The other two modes will lead to a change in the dipole moment and are infrared active.

P8.16) The force constants for F_2 and I_2 are 470. and 172 N m^{-1}, respectively. Calculate the ratio of the vibrational state populations n_1/n_0 and n_2/n_0 at $T = 300.$ and at 1000. K.

The vibrational energy is given by $E_n = \left(n + \dfrac{1}{2}\right)h\nu = \left(n + \dfrac{1}{2}\right)h\sqrt{\dfrac{k}{\mu}}$

For F_2

$$E_0 = \frac{1}{2}\hbar\sqrt{\frac{k}{\mu}} = \frac{1}{2}\times 1.055\times 10^{-34}\ \text{J s}\times\sqrt{\frac{470.\ \text{N m}^{-1}}{\dfrac{18.994\ \text{amu}\times 18.994\ \text{amu}}{18.994\ \text{amu}+18.994\ \text{amu}}\times 1.661\times 10^{-27}\ \text{kg amu}^{-1}}}$$

$= 9.11\times 10^{-21}\ \text{J}$

$$E_1 = \frac{3}{2}\hbar\sqrt{\frac{k}{\mu}} = \frac{3}{2}\times 1.055\times 10^{-34}\ \text{J s}\times\sqrt{\frac{470.\ \text{N m}^{-1}}{\dfrac{18.994\ \text{amu}\times 18.994\ \text{amu}}{18.994\ \text{amu}+18.994\ \text{amu}}\times 1.661\times 10^{-27}\ \text{kg amu}^{-1}}}$$

$= 2.73\times 10^{-20}\ \text{J}$

$$E_2 = \frac{5}{2}\hbar\sqrt{\frac{k}{\mu}} = \frac{5}{2}\times 1.055\times 10^{-34}\ \text{J s}\times\sqrt{\frac{470.\ \text{N m}^{-1}}{\dfrac{18.994\ \text{amu}\times 18.994\ \text{amu}}{18.994\ \text{amu}+18.994\ \text{amu}}\times 1.661\times 10^{-27}\ \text{kg amu}^{-1}}}$$

$= 4.55\times 10^{-20}\ \text{J}$

For I_2

$$E_0 = \frac{1}{2}\hbar\sqrt{\frac{k}{\mu}} = \frac{1}{2}\times 1.055\times 10^{-34}\ \text{J s}\times\sqrt{\dfrac{172\ \text{N m}^{-1}}{\dfrac{126.9045\ \text{amu}\times 126.9045\ \text{amu}}{126.9045\ \text{amu}+126.9045\ \text{amu}}\times 1.661\times 10^{-27}\ \text{kg amu}^{-1}}}$$

$$= 2.13\times 10^{-21}\ \text{J}$$

$$E_1 = \frac{3}{2}\hbar\sqrt{\frac{k}{\mu}} = \frac{3}{2}\times 1.055\times 10^{-34}\ \text{J s}\times\sqrt{\dfrac{172\ \text{N m}^{-1}}{\dfrac{126.9045\ \text{amu}\times 126.9045\ \text{amu}}{126.9045\ \text{amu}+126.9045\ \text{amu}}\times 1.661\times 10^{-27}\ \text{kg amu}^{-1}}}$$

$$= 6.39\times 10^{-21}\ \text{J}$$

$$E_2 = \frac{5}{2}\hbar\sqrt{\frac{k}{\mu}} = \frac{5}{2}\times 1.055\times 10^{-34}\ \text{J s}\times\sqrt{\dfrac{172\ \text{N m}^{-1}}{\dfrac{126.9045\ \text{amu}\times 126.9045\ \text{amu}}{126.9045\ \text{amu}+126.9045\ \text{amu}}\times 1.661\times 10^{-27}\ \text{kg amu}^{-1}}}$$

$$= 1.07\times 10^{-20}\ \text{J}$$

For F_2 at 300. K, $\dfrac{n_1}{n_0} = e^{-\frac{E_1-E_0}{kT}} = e^{-\frac{(2.73-0.911)\times 10^{-20}\text{J}}{1.381\times 10^{-23}\ \text{J K}^{-1}\times 300.\ \text{K}}} = 0.0123$

For F_2 at 1000. K, $\dfrac{n_1}{n_0} = e^{-\frac{E_1-E_0}{kT}} = e^{-\frac{(1.31-0.437)\times 10^{-19}\text{J}}{1.381\times 10^{-23}\ \text{J K}^{-1}\times 1000.\ \text{K}}} = 0.267$

For F_2 at 300. K, $\dfrac{n_2}{n_0} = e^{-\frac{E_2-E_0}{kT}} = e^{-\frac{(4.55-0.915)\times 10^{-20}\text{J}}{1.381\times 10^{-23}\ \text{J K}^{-1}\times 300.\ \text{K}}} = 1.52\times 10^{-4}$

For F_2 at 1000. K, $\dfrac{n_2}{n_0} = e^{-\frac{E_2-E_0}{kT}} = e^{-\frac{(4.55-0.915)\times 10^{-20}\text{J}}{1.381\times 10^{-23}\ \text{J K}^{-1}\times 1000.\ \text{K}}} = 0.0715$

For I_2 at 300. K, $\dfrac{n_1}{n_0} = e^{-\frac{E_1-E_0}{kT}} = e^{-\frac{(6.39-2.13)\times 10^{-21}\text{J}}{1.381\times 10^{-23}\ \text{J K}^{-1}\times 300.\ \text{K}}} = 0.357$

For I_2 at 1000. K, $\dfrac{n_1}{n_0} = e^{-\frac{E_1-E_0}{kT}} = e^{-\frac{(6.39-2.13)\times 10^{-21}\text{J}}{1.381\times 10^{-23}\ \text{J K}^{-1}\times 1000.\ \text{K}}} = 0.734$

For I_2 at 300. K, $\dfrac{n_2}{n_0} = e^{-\frac{E_2-E_0}{kT}} = e^{-\frac{(1.07-0.213)\times 10^{-20}\text{J}}{1.381\times 10^{-23}\ \text{J K}^{-1}\times 300.\ \text{K}}} = 0.127$

For I_2 at 1000. K, $\dfrac{n_2}{n_0} = e^{-\frac{E_2-E_0}{kT}} = e^{-\frac{(1.07-0.213)\times 10^{-20}\text{J}}{1.381\times 10^{-23}\ \text{J K}^{-1}\times 1000.\ \text{K}}} = 0.539$

P8.19) Show that the Morse potential approaches the harmonic potential for small values of the vibrational amplitude. (*Hint:* Expand the Morse potential in a Taylor-Maclaurin series.)

$$V(R) = D_e \left[1 - e^{-\alpha(R-R_e)} \right]^2$$

Expanding in a Taylor-Maclaurin series and keeping only the first term,

$$V(R) = D_e \left[1 - \left[e^{-\alpha(R-R_e)} \right]_{R=R_e} - \left[\frac{\partial e^{-\alpha(R-R_e)}}{\partial(R-R_e)} \right]_{R=R_e} (R-R_e) \right]^2$$

$$= D_e \left[1 - 1 - \alpha(R-R_e) \right]^2 = D_e \alpha^2 (R-R_e)^2$$

P8.20) The rotational constant for $^{127}I^{35}Cl$ determined from microwave spectroscopy is 0.1141619 cm$^-$^{1}. Calculate the bond length in $^{127}I^{35}Cl$ to the maximum number of significant figures consistent with this information.

$$B = \frac{h}{8\pi^2 \mu r_0^2}; \quad r_0 = \sqrt{\frac{h}{8\pi^2 \mu B}}$$

$$r_0 = \sqrt{\frac{6.6260755 \times 10^{-34} \text{ J s}}{8\pi^2 \times \frac{126.904473 \text{ amu} \times 34.9688 \text{ amu}}{(126.904473 \text{ amu} + 34.9688 \text{ amu})} \times 1.6605402 \times 10^{-27} \text{ kg amu}^{-1} \times 0.1141619 \text{ cm}^{-1} \times 2.99792458 \times 10^{10} \text{ cm s}^{-1}}}$$

$$r_0 = 2.32085 \times 10^{-10} \text{ m}$$

P8.23) The fundamental vibrational frequencies for $^1H^{19}F$ and $^2D^{19}F$ are 4138.52 and 2998.25 cm^{-1}, respectively, and D_e for both molecules is 5.86 eV. What is the difference in the bond energy of the two molecules?

$$\left(D_e - \frac{1}{2} hc\tilde{v} \right)_{HF} - \left(D_e - \frac{1}{2} hc\tilde{v} \right)_{DF} = \frac{1}{2} hc\left(\tilde{v}_{HF} - \tilde{v}_{DF} \right)$$

$$= \frac{1}{2} \times 6.626 \times 10^{-34} \text{ J s} \times 3.00 \times 10^{10} \text{ cm s}^{-1} \times \left(4138.52 \text{ cm}^{-1} - 2998.25 \text{ cm}^{-1} \right) = 1.133 \times 10^{-20} \text{ J}$$

P8.27) Fill in the missing step in the derivation that led to the calculation of the spectral line shape in Figure 8.22. Starting from

$$a_2(t) = \mu_x^{21} \frac{E_0}{2} \left(\frac{1 - e^{\frac{i}{\hbar}(E_2-E_1+hv)t}}{E_2 - E_1 + hv} + \frac{1 - e^{-\frac{i}{\hbar}(E_2-E_1-hv)t}}{E_2 - E_1 - hv} \right)$$

and neglecting the first term in the parentheses, show that

$$a_2^*(t)a_2(t) = E_0^2 \left[\mu_x^{21} \right]^2 \frac{\sin^2[(E_2 - E_1 - hv)t/2\hbar]}{(E_2 - E_1 - hv)^2}$$

$$a_2^*(t)a_2(t)=\left[\mu_z^{21}\right]^2 E_0^2\left(\frac{1-e^{+\frac{i}{\hbar}(E_2-E_1-hv)t}}{E_2-E_1-hv}\right)\left(\frac{1-e^{-\frac{i}{\hbar}(E_2-E_1-hv)t}}{E_2-E_1-hv}\right)$$

$$=\left[\mu_z^{21}\right]^2 E_0^2\frac{\left(2-2\cos\left[(E_2-E_1-hv)\frac{t}{\hbar}\right]\right)}{\left(E_2-E_1-hv\right)^2}$$

Using the identity $1-\cos x=1-\cos\left(\frac{x}{2}+\frac{x}{2}\right)=\cos^2\frac{x}{2}+\sin^2\frac{x}{2}-\left(\cos^2\frac{x}{2}-\sin^2\frac{x}{2}\right)=2\sin^2\frac{x}{2}$

$$a_2^*(t)a_2(t)=\left[\mu_z^{21}\right]^2\frac{E_0^2}{4}\left(\frac{2-2\cos\left[(E_2-E_1-hv)\frac{t}{\hbar}\right]}{\left(E_2-E_1-hv\right)^2}\right)=\left[\mu_z^{21}\right]^2 E_0^2\frac{\sin^2\left[(E_2-E_1-hv)\frac{t}{2\hbar}\right]}{\left(E_2-E_1-hv\right)^2}$$

P8.30) A strong absorption band in the infrared region of the electromagnetic spectrum is observed at $\tilde{v}=2649$ cm^{-1} for H^{81}Br. Assuming that the harmonic potential applies, calculate the fundamental frequency v in units of inverse seconds, the vibrational period in seconds, and the zero point energy for the molecule in joules and electron-volts.

$$v=\tilde{v}c=2649 \text{ cm}^{-1}\times 3.00\times 10^{10} \text{ cm s}^{-1}=7.94\times 10^{13} \text{ s}^{-1}$$

$$T=\frac{1}{v}=\frac{1}{7.94\times 10^{13} \text{ s}^{-1}}=1.26\times 10^{-14} \text{ s}$$

$$E=\frac{1}{2}hv=\frac{1}{2}\times 6.626\times 10^{-34} \text{ J s}\times 7.94\times 10^{13} \text{ s}^{-1}=2.63\times 10^{-20} \text{ J}\times\frac{6.241\times 10^{18} \text{ eV}}{\text{J}}=0.164 \text{ eV}$$

P8.31) The spacing between lines in the pure rotational spectrum of ^{7}Li^{1}H is 4.505×10^{11} s^{-1}. Calculate the bond length of this molecule.

$$B = \frac{\nu}{2c} = \frac{4.505 \times 10^{11} \text{ s}^{-1}}{2 \times 2.998 \times 10^{10} \text{ cm s}^{-1}} = 7.513 \text{ cm}^{-1}$$

$$r_0 = \sqrt{\frac{h}{8\pi^2 \mu c B}}$$

$$=$$

$$\sqrt{8\pi^2 \times \frac{1.0078 \text{ amu} \times 7.0160 \text{ amu}}{1.0078 \text{ amu} + 7.0160 \text{ amu}} \times 1.6605 \times 10^{-27} \text{ kg amu}^{-1} \times 2.998 \times 10^{10} \text{ cm s}^{-1} \times 7.513 \text{ cm}^{-1}}$$

Numerator: 6.626×10^{-34} J s

$$r_0 = 1.596 \times 10^{-10} \text{ m}$$

P8.33) Calculate the moment of inertia, the magnitude of the rotational angular momentum, and the energy in the $J = 12$ rotational state for $^1H_2$ in which the bond length of $^1H_2$ is 74.2 pm.

$$I = \mu r_0^2 = \frac{1.007825^2 \text{ amu}^2}{2 \times 1.007825 \text{ amu}} \times 1.6605402 \times 10^{-27} \text{ kg amu}^{-1} \times \left(74.2 \times 10^{-12} \text{ m}\right)^2$$

$$= 4.605 \times 10^{-48} \text{ kg m}^2$$

$$|J| = \sqrt{J(J+1)}\hbar = \sqrt{12 \times 13} \times 1.0554 \times 10^{-34} \text{ J s} = 1.3177 \times 10^{-33} \text{ J s}$$

$$= 1.3177 \times 10^{-33} \text{ kg m s}^{-2}$$

$$E_J = \frac{J(J+1)\hbar^2}{2I} = \frac{12 \times 13 \times \left(1.0554 \times 10^{-34} \text{ J s}\right)^2}{2 \times 4.605 \times 10^{-48} \text{ kg m}^2} = 1.89 \times 10^{-19} \text{ J}$$

P8.38) 50.% of the light incident on a 10.0 mm thick piece of fused silica quartz glass passes through the glass. What percentage of the light will pass through a 20.0 mm thick piece of the same glass?

$$\ln\left[\frac{I(\lambda)}{I_0(\lambda)}\right] = -\varepsilon(\lambda)l; \quad \varepsilon = -\frac{\ln\left[\dfrac{I(\lambda)}{I_0(\lambda)}\right]}{l}$$

$$\varepsilon = -\frac{\ln[0.50]}{1.00 \text{ cm}^{-1}} = 0.693 \text{ cm}^{-1}$$

$$\ln\left[\frac{I(\lambda)}{I_0(\lambda)}\right] = -0.693 \text{ cm}^{-1} \times 2.00 \text{ cm} = -1.3863$$

$$\frac{I(\lambda)}{I_0(\lambda)} = 0.250$$

25% of the light will pass through the glass

P8.41) An infrared absorption spectrum of an organic compound is shown in the following figure. Use the characteristic group frequencies listed in Section 8.5 to decide whether this compound is more likely to be hexene, hexane, or hexanol.

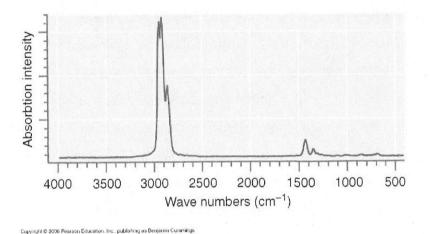

The group of peaks near 2900 cm^{-1} is due to CH_3 and CH_2 stretching vibrations and the small peak near 1400 cm^{-1} is due to a CH_3 umbrella bending mode. These peaks are consistent with the compound being hexane. Hexene should show a strong peak near 1650 cm^{-1} and hexanol should show a strong peak near 3400 cm^{-1}. Because these peaks are absent, these compounds can be ruled out.

Computational Problems

Before solving the computational problems, it is recommended that students work through Tutorials 1–3 under the Help menus in Spartan Student Edition to gain familiarity with the program.

Computational Problem 8.1 Build structures for the gas-phase (a) hydrogen fluoride ($^1H^{19}F$), (b) hydrogen chloride ($^1H^{35}Cl$), (c) carbon monoxide ($^{12}C^{16}O$), and (d) sodium chloride ($^{23}Na^{35}Cl$) molecules. (For Spartan, these are the default isotopic masses.) Calculate the equilibrium geometry and the IR spectrum using the B3LYP method with the 6-311+G** basis set.

a. Compare your result for the vibrational frequency with the experimental value listed in Table 8.3. What is the relative error in the calculation?

b. Calculate the force constant from the vibrational frequency and reduced mass. Determine the relative error using the experimental value in Table 8.3.

c. Calculate the values for the rotational constant, *B*, using the calculated bond length. Determine the relative error using the experimental value in Table 8.3.

Create each one as a new molecule within the same file using the "New Molecule" command in the File menu.

Your solution to this problem should answer the questions in bold type in Steps 5 and 10.

Step 1: Create a new file, and build structures for hydrogen fluoride ($^1H^{19}F$), hydrogen chloride ($^1H^{35}Cl$), carbon monoxide ($^{12}C^{16}O$), and sodium chloride ($^{23}Na^{35}Cl$, just the diatomic, gas-phase species, not the crystal structure!). All of the isotope values are the defaults, so you don't need to worry about changing them. Create each species as a new molecule within the same file using the "New Molecule" command in the File menu. Remember to minimize your structures using molecular mechanics before starting the next step; good initial geometries help keep calculation times short.

Note that for NaCl and CO, you will have to use the inorganic model kit, which is accessed by clicking the "Inorganic" tab on the model kit. To give the atoms only one valence, click the "-X" button. Verify that CO has a triple bond by going to Display > Properties, then clicking on the C-O bond while in view mode. If it is not listed as a triple bond, change its type to triple. While the displayed bond type is not relevant for the QM calculations, it tells the molecular mechanics minimizer to use the right parameters, and give you a better initial geometry.

Step 2: Go to Setup → Calculations. Set the calculation type to "Equilibrium Geometry" and the method to **B3LYP/6-311+G****. The meaning of the method (Density Functional Theory with B3LYP functional) and basis set (6-311+G**) will be explained later in the course, consider this to be a "black box" for now. "Total Charge" should be neutral, "Spin Multiplicity" should be 1, "Global Calculations" should be on, and the "Calculate IR Spectrum" box should be checked. Click "Submit." Your calculations will start immediately.

Since you are using a large set of basis functions (this will be covered later in the course), and since the computational engine is calculating the normal modes for the molecules as well as the energies, the calculations may take several minutes.

Step 3: Once your calculations complete, go to Display → Spectra. You should see the frequency of the normal mode for each molecule, in wave numbers (cm^{-1}). These can be converted to Hz by multiplying by the speed of light in cm/s. You can also draw the spectrum by clicking "Draw Spectrum." If the peak for one or more of the molecules does not appear, you will need to change the scale of the x-axis of the graph. Make sure the "Properties" window is open, then click on the x-axis of the graph while in view mode, and change the viewable range to 100 to 4000 cm^{-1}. Scroll through the individual molecules in the file using the control in the left lower portion of the window.

Step 4: Record the normal mode frequencies for each molecule. Since you will be manipulating these in the next few steps, a spreadsheet may be helpful. Also, measure and record the bond lengths.

Step 5: On your spreadsheet, calculate the reduced mass μ for each molecule, using values taken from the inside back cover of the textbook for the mass in atomic mass units; these are single-isotope species, not a statistical average of isotopes. Sodium is not included in that table; use a mass of 22.9897 amu for Na. Convert the reduced masses into kilograms.

Step 6: Using the normal modes and the reduced masses, calculate the force constant for each molecule using the following equation derived from the harmonic oscillator. Make sure you are using the correct units! Note that this is just a rearrangement of Equation 8.7 from your textbook. Make a table with the bond length, reduced mass, normal mode, and force constant for each molecule, **leaving one extra line for step #9.**

$$k = 4\pi^2 v^2 \mu$$

Step 7: Using the bond lengths and force constants, **qualitatively** estimate the **relative** strength of the bonds in the four species. Is this congruent with your chemical intuition about how strong these bonds should be?

Step 8: Go to http://webbook.nist.gov, (You can access the webbook through the UW library when you log on as member of the UW community) should have a link called "Constants of Diatomic Molecules." Access that data, and obtain the frequency of the normal mode v_e for each species, **in the X$^1\Sigma$ (ground) state**. Be very careful that you pick the correct state. *Compare these values with your calculated values.*

Step 9: The calculated normal mode for deuterium chloride is at 2129.97 cm^{-1}. Calculate the reduced mass and force constant for DCl, and add to your table. *How do the force constants for HCl and DCl compare with each other?*

Step 10: Your calculation can also yield some useful thermodynamic properties. Go to HCl and open up the output, under the display menu. Find the total energy, which is in another potentially unfamiliar unit, the Hartree. Using the total energy for HCl, and the following values for the total energy of H and Cl, calculate the bond dissociation energy.

Cl: -460.166854

H: -0.5021559

Convert this to kJ/mol using the conversion factor 2625.5 kJ/mol per Hartree. *How does it compare with the experimental bond dissociation energy of 427.71 kJ/mol?*

Next, scroll down to the bottom of the output, where you should find a table containing thermodynamic properties. Find and record the total entropy. Next, go back to NIST Webbook, and on the page for HCl, click on "Gas Phase Thermochemistry Data." Find a measurement of the total entropy. *How does your calculated entropy compare with the experimental data?*

Computational Problem 8.2: Calculate the bond energy in gaseous (a) hydrogen fluoride ($^1H^{19}F$), (b) hydrogen chloride ($^1H^{35}Cl$), (c) carbon monoxide ($^{12}C^{16}O$), and (d) sodium chloride ($^{23}Na^{35}Cl$) molecules by comparing the total energies of the species in the dissociation reactions. Use the B3LYP method with the 6-31G* basis set. Determine the relative error of the calculation using the experimental value in Table 8.3.

Step 1: Create a new file, and build structures for hydrogen fluoride ($^1H^{19}F$), hydrogen chloride ($^1H^{35}Cl$), carbon monoxide ($^{12}C^{16}O$), and sodium chloride ($^{23}Na^{35}Cl$, just the diatomic, gas-phase species, not the crystal structure!). Build structures for all the atoms except H for which you know the exact energy. All of the isotope values are the defaults, so you don't need to worry about changing them. Create each species as a new molecule within the same file using the "New Molecule" command in the File menu. Remember to minimize the molecular structures using molecular mechanics before starting the next step; good initial geometries help keep calculation times short.

Note that for NaCl and CO, you will have to use the inorganic model kit, which is accessed by clicking the "Inorganic" tab on the model kit. To give the atoms only one valence, click the "-X" button. Verify that CO has a triple bond by going to Display > Properties, then clicking on the C-O bond while in view mode. If it is not listed as a triple bond, change its type to triple. While the displayed bond type is not relevant for the QM calculations, it tells the molecular mechanics minimizer to use the right parameters, and give you a better initial geometry.

Step 2: Go to Setup → Calculations. Set the calculation type to "Equilibrium Geometry" and the method to **B3LYP/6-311G**** for the molecules and "Energy" for the atoms except for HCl and NaCl. Use the **6-31G*** basis set for these molecules Cl(g) and Na(g). The meaning of the method (Density Functional Theory with B3LYP functional) and basis set (6-31G*) will be explained later in the course, consider this to be a "black box" for now. "Total Charge" is neutral, and you must enter the appropriate number of unpaired electrons for each species. "Global Calculations" should not be on because the number of unpaired electrons is not the same for all species. Click "Submit." Your calculations will start immediately.

Step 3: When the calculation has finished, obtain the total energy for a species by going to Display →

Output. Scroll through the individual species in the file using the control in the left lower portion of

window. Note that the energies are in atomic units where 1 hartree = 2625.5 kJ mol^{-1} = 27.2114 eV.

Record (copy and paste) the energies in a spreadsheet.

Step 4: Calculate the change in total energy for the reaction AB(g) → A(g) + B(g) using the energies

obtained in your calculations. Determine the relative error of the calculation using the experimental

values in Table 8.3 and find a reference value for the NaCl bond energy (the NaCl molecule and not a

NaCl crystal).

Computational Problem 8.3: Build structures for the gas-phase (a) NF_3, (b) PCl_3, and (c) SO_3

molecules. Calculate the equilibrium geometry and the IR spectrum using the B3LYP method with the

6-31G* basis set. Animate the vibrational normal modes and classify them as symmetrical stretch,

symmetrical deformation, degenerate stretch, and degenerate deformation

Step 1: Create a new file, and build structures for $^{14}N^{19}F_3$, $^{31}P^{35}Cl_3$, and $^{32}S^{16}O_3$. All of the isotope

values are the defaults, so you don't need to worry about changing them. Create each species as a new

molecule within the same file using the "New Molecule" command in the File menu. Use the inorganic

model kit, which is accessed by clicking the "Inorganic" tab on the model kit. Remember to minimize

your structures using molecular mechanics before starting the next step; good initial geometries help

keep calculation times short.

Step 2: Go to Setup → Calculations. Set the calculation type to "Equilibrium Geometry" and the

method to **B3LYP/6-31G***. The meaning of the method (Density Functional Theory with B3LYP

functional) and basis set (6-31G*) will be explained later in the course, consider this to be a "black box"

for now. "Total Charge" should be neutral. Enter the appropriate number of unpaired electrons. "Global

Calculations" should be checked, and **the "Calculate IR Spectrum" box should be checked.** Click

"Submit." Your calculations will start immediately.

Step 3: Once your calculations complete, go to Display → Spectra. You should see the frequency of the

normal modes for each molecule, in wave numbers (cm^{-1}). Note the normal mode frequencies for each

molecule as well as their degeneracies. How many normal modes should each molecule exhibit? If one

or more of the peaks does not appear, you will need to change the scale of the x-axis of the graph. Make

sure the "Properties" window is open, then click on the graph while in view mode, and change the

viewable range to include all normal mode frequencies. Scroll through the individual molecules in the

file using the control in the left lower portion of the window.

Step 4: Make a table showing the normal mode frequencies, the degeneracy (number of modes with the same frequency), and intensity in columns 1–3 for each molecule. How many peaks have sufficient intensity that they are easily observable? If any peaks have zero intensity, explain why. Classify the modes as symmetrical stretch, symmetrical deformation, degenerate stretch, and degenerate deformation. A stretching mode is symmetric if all bonds are being stretched simultaneously, and degenerate if two or more stretching modes have the same frequency. The same classification holds for a deformation, in which the bond angle, rather than the bond length, changes. Compare the frequencies of equivalent normal modes in NF_3 and PCl_3. What factors might lead to the observed difference in frequency?

Computational Problem 8.4: Build structures for the gas-phase (a) F_2CO, (b) Cl_2CO, and (c) O_2NF molecules of the structural form X_2YZ. Calculate the equilibrium geometry and the IR spectrum using the B3LYP method with the 6-311+G**basis set. Animate the vibrational normal modes and classify them as symmetric Y-Z stretch, YX_2 scissors, antisymmetric X-Y stretch, YX_2 rock, and $Y-X_2$ wag.

Step 1: Create a new file, and build structures for (a) $^{19}F_2{}^{12}C^{16}O$, (b) $^{35}Cl_2{}^{12}C^{16}O$, and (c) $^{16}O_2{}^{14}N^{19}F$. All of the isotope values are the defaults, so you don't need to worry about changing them. Create each species as a new molecule within the same file using the "New Molecule" command in the File menu. Use the inorganic model kit, which is accessed by clicking the "Inorganic" tab on the model kit. Remember to minimize your structures using molecular mechanics before starting the next step; good initial geometries help keep calculation times short.

Step 2: Go to Setup → Calculations. Set the calculation type to "Equilibrium Geometry" and the method to **B3LYP/6-31G***. The meaning of the method (Density Functional Theory with B3LYP functional) and basis set (6-31G*) will be explained later in the course, consider this to be a "black box" for now. "Total Charge" should be neutral. Enter the appropriate number of unpaired electrons. "Global Calculations" should be checked, and **the "Calculate IR Spectrum" box should be checked.** Click "Submit." Your calculations will start immediately.

Step 3: Once your calculations complete, go to Display → Spectra. You should see the frequency of the normal modes for each molecule, in wave numbers (cm^{-1}). Note the normal mode frequencies for each molecule as well as their degeneracies. How many normal modes should each molecule exhibit? If one or more of the peaks does not appear, you will need to change the scale of the x-axis of the graph. Make sure the "Properties" window is open, then click on the graph while in view mode, and change the viewable range to include all normal mode frequencies. Scroll through the individual molecules in the file using the control in the left lower portion of the window.

Step 4: Make a table showing the normal mode frequencies, the degeneracy (number of modes with the same frequency), and intensity in columns 1–3 for each molecule. How many peaks have sufficient intensity that they are easily observable? If any peaks have zero intensity, explain why. Classify the modes as symmetric Y-Z stretch, YX_2 scissors, antisymmetric X-Y stretch, YX_2 rock, and Y-X_2 wag. A stretching mode is symmetric if all bonds are being stretched simultaneously, and degenerate if two or more stretching modes have the same frequency. An out of plane deformation is referred to as a wag. Rocking and scissors modes correspond to their names. Compare the frequencies of equivalent normal modes in F_2CO and Cl_2CO. What factors might lead to the observed difference in frequency? Is the order of the modes symmetric Y-Z stretch, YX_2 scissors, antisymmetric X-Y stretch, YX_2 rock, and Y-X_2 wag the same for all molecules in terms of frequency?

Computational Problem 8.5: Build structures for the bent gas-phase (a) HOF, (b) ClOO, and (c) HSO molecules of the structural form XYZ. Calculate the equilibrium geometry and the IR spectrum using the B3LYP method with the basis set. Animate the vibrational normal modes and classify them as Y-Z stretch, X-Y stretch, and X-Y-Z bend.

Step 1: Create a new file, and build structures for $^1H^{16}O^{19}F$, (b) $^{35}Cl^{16}O^{16}O$, and (c) $^1H^{32}S^{16}O$. All of the isotope values are the defaults, so you don't need to worry about changing them. Create each species as a new molecule within the same file using the "New Molecule" command in the File menu. Use the inorganic model kit, which is accessed by clicking the "Inorganic" tab on the model kit. Remember to minimize your structures using molecular mechanics before starting the next step; good initial geometries help keep calculation times short.

Step 2: Go to Setup → Calculations. Set the calculation type to "Equilibrium Geometry" and the method to **B3LYP/6-31G***. The meaning of the method (Density Functional Theory with B3LYP functional) and basis set (6-31G*) will be explained later in the course, consider this to be a "black box" for now. "Total Charge" should be neutral. Enter the appropriate number of unpaired electrons. "Global Calculations" should not be checked, and **the "Calculate IR Spectrum" box should be checked.** Click "Submit." Your calculations will start immediately.

Step 3: Once your calculations complete, go to Display → Spectra. You should see the frequency of the normal modes for each molecule, in wave numbers (cm^{-1}). Note the normal mode frequencies for each molecule as well as their degeneracies. How many normal modes should each molecule exhibit? If one or more of the peaks does not appear, you will need to change the scale of the x-axis of the graph. Make sure the "Properties" window is open, then click on the graph while in view mode, and change the viewable range to include all normal mode frequencies. Scroll through the individual molecules in the file using the control in the left lower portion of the window.

Step 4: Make a table showing the normal mode frequencies, the degeneracy (number of modes with the same frequency), and intensity in columns 1–3 for each molecule. How many peaks have sufficient intensity that they are easily observable? If any peaks have zero intensity, explain why. Classify the modes as Y-Z stretch, X-Y stretch, and X-Y-Z bend.

Chapter 9: The Hydrogen Atom

P9.1) Calculate the wave number corresponding to the most and least energetic spectral lines in the Lyman, Balmer, and Paschen series for the hydrogen atom.

$$\tilde{v} = 109{,}678 \left(\frac{1}{n_{initial}^2} - \frac{1}{n_{final}^2} \right) cm^{-1} \text{ and the most energetic line corresponds to } n \to \infty.$$

$n_{initial}$ is 1, 2, and 3 for the Lyman, Balmer, and Paschen series, respectively.

$\tilde{v} = 109{,}678 \text{ cm}^{-1}$ for the Lyman series,

$$\tilde{v} = 109{,}678 \left(\frac{1}{4} \right) cm^{-1} = 27419.3 \, cm^{-1} \text{ for the Balmer series, and}$$

$$\tilde{v} = 109{,}678 \left(\frac{1}{9} \right) cm^{-1} = 12186.3 \, cm^{-1} \text{ for the Paschen series.}$$

The least energetic transition corresponds to $n_{initial} \to n_{initial} + 1$.

$$\tilde{v} = 109{,}678 \left(1 - \frac{1}{4} \right) cm^{-1} = 82258.5 \, cm^{-1},$$

$$\tilde{v} = 109{,}678 \left(\frac{1}{4} - \frac{1}{9} \right) cm^{-1} = 15233.1 \, cm^{-1}, \text{ and}$$

$$\tilde{v} = 109{,}678 \left(\frac{1}{9} - \frac{1}{16} \right) cm^{-1} = 5331.57 \, cm^{-1} \text{ for the Lyman, Balmer, and Paschen series, respectively.}$$

P9.5) Using the result of Problem P9.13, calculate the probability that the $1s$ electron for H will be found between $r = 0$ and $r = a_0$.

Let $u = r^2$ and $dv = e^{-\frac{r}{\alpha}} dr$

$$\int r^2 e^{-\frac{r}{\alpha}} dr = -\alpha r^2 e^{-\frac{r}{\alpha}} + 2\alpha \int r e^{-\frac{r}{\alpha}} dr$$

Integrating by parts again,

$$\int r^2 e^{-\frac{r}{\alpha}} dr = -\alpha r^2 e^{-\frac{r}{\alpha}} + 2\alpha \left(-\alpha r e^{-\frac{r}{\alpha}} + \alpha \int e^{-\frac{r}{\alpha}} dr \right) = -\alpha r^2 e^{-\frac{r}{\alpha}} - 2\alpha \left(\alpha r e^{-\frac{r}{\alpha}} - \alpha^2 e^{-\frac{r}{\alpha}} \right)$$

$$\int r^2 e^{-\frac{r}{\alpha}} dr = e^{-\frac{r}{\alpha}} \left(-\alpha r^2 - 2\alpha^2 r - 2\alpha^3 \right)$$

$$P = 1 - e^{-2r/a_0} - \frac{2r}{a_0}\left(1 + \frac{r}{a_0}\right)e^{-2r/a_0}$$

for $r = a_0$

P9.6) Using the result of Problem P9.13, calculate the probability of finding the electron in the $1s$ state outside a sphere of radius $0.5a_0$, $3a_0$, and $5a_0$.

The probability of finding the electron inside the sphere of radius r is $1 - e^{-\frac{2r}{a_0}} - \frac{2r}{a_0}\left(1 + \frac{r}{a_0}\right)e^{-\frac{2r}{a_0}}$. The

probability of finding it outside the sphere of radius r is

$$1 - \left[1 - e^{-\frac{2r}{a_0}} - \frac{2r}{a_0}\left(1 + \frac{r}{a_0}\right)e^{-\frac{2r}{a_0}}\right] = e^{-\frac{2r}{a_0}} + \frac{2r}{a_0}\left(1 + \frac{r}{a_0}\right)e^{-\frac{2r}{a_0}}.$$ Evaluating this function at

$0.5a_0$, $3a_0$, and $5a_0$ gives 0.920, 0.0620, and 2.77×10^{-3}, respectively.

P9.12) As the principal quantum number n increases, the electron is more likely to be found far from the nucleus. It can be shown that for H and for ions with only one electron such as He$^+$,

$$\langle r \rangle_{nl} = \frac{n^2 a_0}{Z}\left[1 + \frac{1}{2}\left(1 - \frac{l(l+1)}{n^2}\right)\right]$$

Calculate the value of n for an s state in the hydrogen atom such that $\langle r \rangle = 1000\, a_0$. Round up to the nearest integer. What is the ionization energy of the H atom in this state in electron-volts? Compare your answer with the ionization energy of the H atom in the ground state.

$$\langle r \rangle_{nl} = \frac{n^2 a_0}{Z}\left[1 + \frac{1}{2}\left(1 - \frac{l(l+1)}{n^2}\right)\right]$$

$$n = \sqrt{\frac{2Z\langle r \rangle_{n0}}{3a_0}} = \sqrt{\frac{2000 a_0}{3a_0}} = 25.82 \approx 26$$

$$I = \frac{Z^2 e^2}{8\pi\, \varepsilon_0\, a_0\, n^2} = \frac{\frac{Z^2}{n^2} \times \left(1.602 \times 10^{-19}\ \text{C}\right)^2}{8\pi \times 8.854 \times 10^{-12}\ \text{J}^{-1}\ \text{C}^2\ \text{m}^{-1} \times 5.292 \times 10^{-11}\ \text{m}} \times \frac{1\text{eV}}{1.602 \times 10^{-19}\ \text{J}}$$

$$I = 13.6039\frac{Z^2}{n^2}\ \text{eV} = 13.6039\frac{1}{n^2}\ \text{eV} \quad \text{for the H atom.}$$

For the ground state, $I = 13.6039$ eV and for $n = 26$, $I \approx 0.0201$ eV.

P9.14) Calculate the expectation value $(r - \langle r \rangle)^2$ if the H atom wave function is $\psi_{100}(r)$.

$$\left\langle (r - \langle r \rangle)^2 \right\rangle = \left\langle r^2 - 2r\langle r \rangle + \langle r \rangle^2 \right\rangle = \left\langle r^2 \right\rangle - 2\langle r \rangle^2 + \langle r \rangle^2 = \left\langle r^2 \right\rangle - \langle r \rangle^2$$

$$\langle r \rangle = \frac{1}{\pi a_0^3} \int_0^{2\pi} d\phi \int_0^{\pi} \sin\theta \, d\theta \int_0^{\infty} r^3 e^{-\frac{2r}{a_0}} dr$$

$$\langle r \rangle = \frac{4}{a_0^3} \int_0^{\infty} r^3 e^{-\frac{2r}{a_0}} dr$$

Using the standard integral $\int_0^{\infty} r^n e^{-\alpha r} = \dfrac{n!}{\alpha^{n+1}}$

$$\langle r \rangle = \frac{4}{a_0^3} \frac{6 a_0^4}{16} = \frac{3}{2} a_0$$

$$\left\langle r^2 \right\rangle = \frac{1}{\pi a_0^3} \int_0^{2\pi} d\phi \int_0^{\pi} \sin\theta \, d\theta \int_0^{\infty} r^4 e^{-\frac{2r}{a_0}} dr$$

$$\left\langle r^2 \right\rangle = \frac{4}{a_0^3} \int_0^{\infty} r^4 e^{-\frac{2r}{a_0}} dr$$

Using the standard integral $\int_0^{\infty} r^n e^{-\alpha r} = \dfrac{n!}{\alpha^{n+1}}$

$$\left\langle r^2 \right\rangle = \frac{4}{a_0^3} \frac{4!}{(2/a_0)^5} = 3(a_0)^2$$

Therefore, $(r - \langle r \rangle)^2 = \left\langle r^2 \right\rangle - \langle r \rangle^2 = 3(a_0)^2 - \left(\frac{3}{2} a_0\right)^2 = \frac{3}{4} a_0^2.$

P9.16) The force acting between the electron and the proton in the H atom is given by

$F = -e^2 / 4\pi\varepsilon_0 r^2$. Calculate the expectation value $\langle F \rangle$ for the $1s$ and $2p_z$ states of the H atom in terms of

e, ε_0, and a_0.

$$\langle F \rangle_{1s} = -\frac{e^2}{4\pi\varepsilon_0} \int \psi^* (\tau) \frac{1}{r^2} \psi(\tau) d\tau$$

$$\langle F \rangle_{1s} = -\frac{e^2}{4\pi\varepsilon_0} \frac{1}{\pi a_0^3} \int_0^{2\pi} d\phi \int_0^{\pi} \sin\theta\, d\theta \int_0^{\infty} \left[e^{-r/a_0} \right] \left(\frac{1}{r^2} \right) \left[e^{-r/a_0} \right] r^2 dr$$

$$\langle F \rangle_{1s} = -\frac{e^2}{4\pi\varepsilon_0} \frac{4}{a_0^3} \int_0^{\infty} e^{-2r/a_0} dr = -\frac{e^2}{4\pi\varepsilon_0} \frac{4}{a_0^3} \left[-\frac{a_0}{2} e^{-2r/a_0} \right]_0^{\infty} = -\frac{e^2}{2\pi\varepsilon_0 a_0^2}$$

$$\langle F \rangle_{2pz} = -\frac{e^2}{4\pi\varepsilon_0} \int \psi^* (\tau) \frac{1}{r^2} \psi(\tau) d\tau$$

$$\langle F \rangle_{2pz} = -\frac{e^2}{4\pi\varepsilon_0} \frac{1}{32\pi a_0^3} \int_0^{2\pi} d\phi \int_0^{\pi} \cos^2\theta \sin\theta\, d\theta \int_0^{\infty} \left(\frac{r}{a_0} \right)^2 \left[e^{-r/a_0} \right] \left(\frac{1}{r^2} \right) r^2 dr$$

$$\langle F \rangle_{2pz} = -\frac{e^2}{4\pi\varepsilon_0} \frac{1}{16 a_0^5} \left[\frac{\cos^3\theta}{3} \right]_0^{\pi} \times \int_0^{\infty} r^2 e^{-r/a_0} dr$$

$$\langle F \rangle_{2pz} = -\frac{e^2}{4\pi\varepsilon_0} \frac{1}{24 a_0^3} \int_0^{\infty} r^2 e^{-r/a_0} dr$$

Using the standard integral $\int_0^{\infty} r^n e^{-\alpha r} = \dfrac{n!}{\alpha^{n+1}}$

$$\langle F \rangle_{2pz} = -\frac{e^2}{4\pi\varepsilon_0} \frac{1}{24 a_0^5} \times 2 a_0^3 = -\frac{e^2}{48\pi\varepsilon_0 a_0^2}$$

P9.17) The d orbitals have the nomenclature $d_{z^2}, d_{xy}, d_{xz}, d_{yz}$, and $d_{x^2-y^2}$. Show how the d orbital

$$\psi_{3d_{yz}}(r,\theta,\phi) = \frac{\sqrt{2}}{81\sqrt{\pi}} \left(\frac{1}{a_0} \right)^{3/2} \frac{r^2}{a_0^2} e^{-r/3a_0} \sin\theta \cos\theta \sin\phi$$

can be written in the form $yzF(r)$.

In spherical coordinates, $x = r\sin\theta\cos\phi$, $y = r\sin\theta\sin\phi$, and $z = r\cos\theta$. Therefore,

$$\psi_{3d_{yz}}(r,\theta,\phi) = \frac{\sqrt{2}}{81\sqrt{\pi}} \left(\frac{1}{a_0} \right)^{3/2} \frac{r^2}{a_0^2} e^{-r/3a_0} \sin\theta \cos\theta \sin\phi$$

$$= \frac{\sqrt{2}}{81\sqrt{\pi}} \left(\frac{1}{a_0} \right)^{3/2} \frac{1}{a_0^2} e^{-r/3a_0} (r\cos\theta)(r\sin\theta\sin\phi) = \frac{\sqrt{2}}{81\sqrt{\pi}} \left(\frac{1}{a_0} \right)^{3/2} \frac{1}{a_0^2} e^{-r/3a_0} (yz)$$

P9.18) Calculate the expectation value of the moment of inertia of the H atom in the 2s and $2p_z$ states in terms of μ and a_0.

$$\langle I \rangle = \langle \mu r^2 \rangle = \mu \frac{1}{32\pi a_0^3} \int_0^{2\pi} d\phi \int_0^{\pi} \sin\theta \, d\theta \int_0^{\infty} r^4 \left(2 - \frac{r}{a_0}\right)^2 e^{-r/a_0} dr$$

$$= \mu \frac{1}{8a_0^3} \int_0^{\infty} \left(4r^4 - \frac{4r^5}{a_0} + \frac{r^6}{a_0^2}\right) e^{-r/a_0} dr = \mu \frac{1}{8a_0^3} \left(4\int_0^{\infty} r^4 \, e^{-r/a_0} dr - \frac{4}{a} \int_0^{\infty} r^5 \, e^{-r/a_0} dr + \frac{1}{a^2}\int_0^{\infty} r^6 \, e^{-r/a_0} dr\right)$$

Using the standard integral $\int_0^{\infty} r^n e^{-\alpha r} = \dfrac{n!}{\alpha^{n+1}}$

$$\langle I \rangle = \mu \frac{1}{8a_0^3} \left(4 \times 4! a_0^5 - \frac{4}{a_0} \times 5! \times a_0^6 + \frac{1}{a_0^2} \times 6! \times a_0^7\right) = 42\mu a_0^2$$

For the $2p_z$ state,

$$\langle I \rangle = \langle \mu r^2 \rangle = \mu \frac{1}{32\pi a_0^3} \int_0^{2\pi} d\phi \int_0^{\pi} \cos^2\theta \sin\theta \, d\theta \int_0^{\infty} r^4 \left(\frac{r}{a_0}\right)^2 e^{-r/a_0} dr$$

$$= \mu \frac{1}{16a_0^5} \left[\frac{\cos^3\theta}{3}\right]_0^{\pi} \times \int_0^{\infty} r^6 e^{-r/a_0} \, dr = \mu \frac{1}{24a_0^5} 6! a_0^7 = 30\mu a_0^2$$

P9.19) The energy levels for ions with a single electron such as He^+, Li^{2+}, and Be^{3+} are given by $E_n = -Z^2 e^2 / 8\pi\varepsilon_0 a_0 n^2, n = 1, 2, 3, 4, \ldots$. Calculate the ionization energies of H, He^+, Li^{2+}, and Be^{3+} in their ground states in units of electron-volts (eV).

The ionization potential is the negative of the orbital energy.

$$I = \frac{Z^2 e^2}{8\pi\varepsilon_0 a_0 n^2} = \frac{\dfrac{Z^2}{n^2} \times \left(1.602 \times 10^{-19} \text{ C}\right)^2}{8\pi \times 8.854 \times 10^{-12} \text{ J}^{-1} \text{ C}^2 \text{ m}^{-1} \times 5.292 \times 10^{-11} \text{ m}} \times \frac{1 \text{ eV}}{1.602 \times 10^{-19} \text{ J}}$$

$$I = 13.60 \frac{Z^2}{n^2} \text{ eV}$$

$$I_H = 13.60 \text{ eV}; \quad I_{He^+} = 54.42 \text{ eV}; \quad I_{Li^{2+}} = 122.4 \text{ eV}; \quad I_{Be^{3+}} = 217.7 \text{ eV}$$

P9.22) Locate the radial and angular nodes in the H orbitals $\psi_{3p_x}(r, \theta, \phi)$ and $\psi_{3p_z}(r, \theta, \phi)$.

$$\psi_{3p_x}(r,\theta,\phi) = \frac{1}{81\sqrt{2\pi}}\left(\frac{1}{a_0}\right)^{3/2}\left(6\frac{r}{a_0}-\frac{r^2}{a_0^2}\right)e^{-r/3a_0}\sin\theta\cos\phi \text{ has radial nodes where } 6\frac{r}{a_0}-\frac{r^2}{a_0^2} \text{ has}$$

zeros. This is at $r = 0$, which does not count as a node, and $r = 6a_0$. The angular nodes correspond to

the angles at which the function is zero, or $\theta = 0°$ and $\phi = 90°$, corresponding to the y-z plane.

$$\psi_{3p_z}(r,\theta,\phi) = \frac{1}{81}\left(\frac{2}{\pi}\right)^{1/2}\left(\frac{1}{a_0}\right)^{3/2}\left(6\frac{r}{a_0}-\frac{r^2}{a_0^2}\right)e^{-r/3a_0}\cos\theta \text{ also has a radial node at } r = 6a_0. \text{ The angular}$$

node is for $\theta = 90°$, corresponding to the x-y plane.

P9.25) Show that the total energy eigenfunctions $\psi_{100}(r)$ and $\psi_{200}(r)$ are orthogonal.

$$\iiint \psi^*_{100}(\tau)\psi_{200}(\tau)d\tau = \frac{1}{\sqrt{32}\,\pi a_0^3}\int\limits_0^{2\pi} d\phi\int\limits_0^{\pi}\sin\theta\,d\theta\int\limits_0^{\infty} r^2 e^{-r/a_0}\left(2-\frac{r}{a_0}\right)e^{-r/2a_0}$$

$$= \frac{1}{\sqrt{2}\,a_0^3}\times\left(\int\limits_0^{\infty} 2r^2 e^{-3r/2a_0}\,dr - \int\limits_0^{\infty}\frac{r^3}{a_0}e^{-3r/2a_0}\,dr\right)$$

$$= \frac{1}{\sqrt{2}\,a_0^3}\times\left(\frac{4}{[3/2a_0]^3} - \frac{6}{a_0[3/2a_0]^4}\right) = \frac{1}{\sqrt{2}\,a_0^3}\times\left(\frac{32a_0^3}{27} - \frac{96a_0^3}{81}\right) = 0$$

Chapter 10: Many-Electron Atoms

P10.1) Is $\psi(1,2) = 1s(1)\alpha(1)1s(2)\beta(2) + 1s(2)\alpha(2)1s(1)\beta(1)$ an eigenfunction of the operator $\hat{S}_z$? If so, what is its eigenvalue M_S?

This problem uses a notation that is not explained in this chapter, but is explained in Chapter 11.

$$\hat{S}_z \left[1s(1)\alpha(1)1s(2)\beta(2) + 1s(2)\alpha(2)1s(1)\beta(1) \right]$$

$$= (\hat{s}_{z1} + \hat{s}_{z2}) \left[1s(1)\alpha(1)1s(2)\beta(2) + 1s(2)\alpha(2)1s(1)\beta(1) \right]$$

$$= \frac{\hbar}{2} \left[1s(1)\alpha(1)1s(2)\beta(2) - 1s(2)\alpha(2)1s(1)\beta(1) \right]$$

$$+ \frac{\hbar}{2} \left[-1s(1)\alpha(1)1s(2)\beta(2) + 1s(2)\alpha(2)1s(1)\beta(1) \right]$$

$$= \frac{\hbar}{2}(0)$$

The function is an eigenfunction of $\hat{S}_z$ with the eigenvalue $M_S = 0$.

P10.4) In this problem you will prove that the ground-state energy for a system obtained using the variational method is greater than the true energy.

a. The approximate wave function Φ can be expanded in the true (but unknown) eigenfunctions ψ_n of the total energy operator in the form $\Phi = \sum_n c_n \psi_n$. Show that by substituting $\Phi = \sum_n c_n \psi_n$ in the equation

$$E = \frac{\int \Phi^* \hat{H} \Phi \, d\tau}{\int \Phi^* \Phi \, d\tau}$$

you obtain the result

$$E = \frac{\displaystyle\sum_n \sum_m \int (c_n^* \psi_n^*) \hat{H} (c_m \psi_m) \, d\tau}{\displaystyle\sum_n \sum_m \int (c_n^* \psi_n^*)(c_m \psi_m) \, d\tau}$$

b. Because the ψ_n are eigenfunctions of $\hat{H}$, they are orthogonal and $\hat{H}\psi_n = E_n\psi_n$. Show that this information allows us to simplify the expression for D from part (a) to

$$E = \frac{\sum_m E_m c_m^* c_m}{\sum_m c_m^* c_m}$$

c. Arrange the terms in the summation such that the first energy is the true ground-state energy E_0 and the energy increases with the summation index m. Why can you conclude that $E - E_0 \geq 0$?

a) $E = \dfrac{\int \Phi^* \hat{H}\Phi \, d\tau}{\int \Phi^* \Phi \, d\tau} = \dfrac{\int \left(\Phi = \sum_n c_n\psi_n\right)^* \hat{H}\left(\Phi = \sum_m c_m\psi_m\right) d\tau}{\int \left(\Phi = \sum_n c_n\psi_n\right)^* \left(\Phi = \sum_m c_m\psi_m\right) d\tau} = \dfrac{\sum_n \sum_m \int \left(c_n^*\psi_n^*\right)\hat{H}\left(c_m\psi_m\right) d\tau}{\sum_n \sum_m \int \left(c_n^*\psi_n^*\right)\left(c_m\psi_m\right) d\tau}$

b) Simplify the previous expression for E from part (a) to $E = \dfrac{\sum_m E_m c_m^* c_m}{\sum_m c_m^* c_m}$

$E = \dfrac{\sum_n \sum_m \int \left(c_n^*\psi_n^*\right)\hat{H}\left(c_m\psi_m\right) d\tau}{\sum_n \sum_m \int \left(c_n^*\psi_n^*\right)\left(c_m\psi_m\right) d\tau} = \dfrac{\sum_n \sum_m E_m \int \left(c_n^*\psi_n^*\right)\left(c_m\psi_m\right) d\tau}{\sum_n \sum_m \int \left(c_n^*\psi_n^*\right)\left(c_m\psi_m\right) d\tau}$

$= \dfrac{\sum_m E_m c_m^* c_m \int \psi_m^*\psi_m \, d\tau}{\sum_m c_m^* c_m \int \psi_m^*\psi_m \, d\tau} = \dfrac{\sum_m E_m c_m^* c_m}{\sum_m c_m^* c_m}$

c)

$E - E_0 = \dfrac{\sum_m E_m c_m^* c_m}{\sum_m c_m^* c_m} - \dfrac{E_0 \sum_m c_m^* c_m}{\sum_m c_m^* c_m} = \dfrac{\sum_m \left(E_m - E_0\right) c_m^* c_m}{\sum_m c_m^* c_m} \geq 0.$ Both $\left(E_m - E_0\right)$ and $c_m^* c_m$ are greater than

zero. Therefore, $E - E_0 \geq 0$.

P10.6) The operator for the square of the total spin of two electrons is $\hat{S}_{total}^2 = (\hat{S}_1 + \hat{S}_2)^2 = \hat{S}_1^2 + \hat{S}_2^2 + 2(\hat{S}_{1x}\hat{S}_{2x} + \hat{S}_{1y}\hat{S}_{2y} + \hat{S}_{1z}\hat{S}_{2z})$. Given that

$$\hat{S}_x \alpha = \frac{\hbar}{2}\beta, \quad \hat{S}_y \alpha = \frac{i\hbar}{2}\beta, \quad \hat{S}_z \alpha = \frac{\hbar}{2}\alpha,$$

$$\hat{S}_x \beta = \frac{\hbar}{2}\alpha, \quad \hat{S}_y \beta = -\frac{i\hbar}{2}\alpha, \quad \hat{S}_z \beta = -\frac{\hbar}{2}\beta,$$

show that $\alpha(1)\,\alpha(2)$ and $\beta(1)\,\beta(2)$ are eigenfunctions of the operator $\hat{S}_{total}^2$. What is the eigenvalue in each case?

This problem uses a notation that is not explained in this chapter, but is explained in Chapter 11

$\hat{S}_{total}^2 \alpha(1)\alpha(2)$

$= \hat{S}_1^2 \alpha(1)\alpha(2) + \hat{S}_2^2 \alpha(1)\alpha(2) + 2\left(\hat{S}_{1x}\hat{S}_{2x}\alpha(1)\alpha(2) + \hat{S}_{1y}\hat{S}_{2y}\alpha(1)\alpha(2) + \hat{S}_{1z}\hat{S}_{2z}\alpha(1)\alpha(2)\right)$

$= \alpha(2)\hat{S}_1^2 \alpha(1) + \alpha(1)\hat{S}_2^2 \alpha(2) + 2\left(\hat{S}_{1x}\alpha(1)\hat{S}_{2x}\alpha(2) + \hat{S}_{1y}\alpha(1)\hat{S}_{2y}\alpha(2) + \hat{S}_{1z}\alpha(1)\hat{S}_{2z}\alpha(2)\right)$

$= \frac{3\hbar^2}{4}\alpha(1)\alpha(2) + \frac{3\hbar^2}{4}\alpha(1)\alpha(2) + 2\left(\hat{S}_{1x}\alpha(1)\hat{S}_{2x}\alpha(2) + \hat{S}_{1y}\alpha(1)\hat{S}_{2y}\alpha(2) + \hat{S}_{1z}\alpha(1)\hat{S}_{2z}\alpha(2)\right)$

$= \frac{3\hbar^2}{4}\alpha(1)\alpha(2) + \frac{3\hbar^2}{4}\alpha(1)\alpha(2) + 2 \times \frac{\hbar}{2}\left(\hat{S}_{1x}\alpha(1)\beta(2) + i\hat{S}_{1y}\alpha(1)\beta(2) + \hat{S}_{1z}\alpha(1)\alpha(2)\right)$

$= \frac{3\hbar^2}{4}\alpha(1)\alpha(2) + \frac{3\hbar^2}{4}\alpha(1)\alpha(2) + 2 \times \left(\frac{\hbar}{2}\right)^2 \left(\beta(1)\beta(2) + i^2\beta(1)\beta(2) + \alpha(1)\alpha(2)\right)$

$= \frac{3\hbar^2}{4}\alpha(1)\alpha(2) + \frac{3\hbar^2}{4}\alpha(1)\alpha(2) + \frac{2\hbar^2}{4}\alpha(1)\alpha(2) = 2\hbar^2 \alpha(1)\alpha(2)$

The eigenvalue is $2\hbar^2$.

$\hat{S}_{total}^2 \beta(1)\beta(2)$

$= \hat{S}_1^2 \beta(1)\beta(2) + \hat{S}_2^2 \beta(1)\beta(2) + 2\left(\hat{S}_{1x}\hat{S}_{2x}\beta(1)\beta(2) + \hat{S}_{1y}\hat{S}_{2y}\beta(1)\beta(2) + \hat{S}_{1z}\hat{S}_{2z}\beta(1)\beta(2)\right)$

$= \alpha(2)\hat{S}_1^2 \beta(1) + \beta(1)\hat{S}_2^2 \beta(2) + 2\left(\hat{S}_{1x}\beta(1)\hat{S}_{2x}\beta(2) + \hat{S}_{1y}\beta(1)\hat{S}_{2y}\beta(2) + \hat{S}_{1z}\beta(1)\hat{S}_{2z}\beta(2)\right)$

$= \frac{3\hbar^2}{4}\beta(1)\beta(2) + \frac{3\hbar^2}{4}\beta(1)\beta(2) + 2\left(\hat{S}_{1x}\beta(1)\hat{S}_{2x}\beta(2) + \hat{S}_{1y}\beta(1)\hat{S}_{2y}\beta(2) + \hat{S}_{1z}\beta(1)\hat{S}_{2z}\beta(2)\right)$

$= \frac{3\hbar^2}{4}\alpha(1)\alpha(2) + \frac{3\hbar^2}{4}\beta(1)\beta(2) + 2 \times \frac{\hbar}{2}\left(\hat{S}_{1x}\beta(1)\alpha(2) - i\hat{S}_{1y}\beta(1)\alpha(2) - \hat{S}_{1z}\beta(1)\beta(2)\right)$

$= \frac{3\hbar^2}{4}\alpha(1)\alpha(2) + \frac{3\hbar^2}{4}\beta(1)\beta(2) + 2 \times \left(\frac{\hbar}{2}\right)^2 \left(\alpha(1)\alpha(2) + i^2\alpha(1)\alpha(2) + \beta(1)\beta(2)\right)$

$= \frac{3\hbar^2}{4}\alpha(1)\alpha(2) + \frac{3\hbar^2}{4}\beta(1)\beta(2) + \frac{2\hbar^2}{4}\beta(1)\beta(2) = 2\hbar^2 \beta(1)\beta(2)$

The eigenvalue is $2\hbar^2$.

P10.11) Write the Slater determinant for the ground-state configuration of Be.

$$\psi_{Be} = \frac{1}{\sqrt{4!}} \begin{vmatrix} 1s(1)\alpha(1) & 1s(1)\beta(1) & 2s(1)\alpha(1) & 2s(1)\beta(1) \\ 1s(2)\alpha(2) & 1s(2)\beta(2) & 2s(2)\alpha(2) & 2s(2)\beta(2) \\ 1s(3)\alpha(3) & 1s(3)\beta(3) & 2s(3)\alpha(3) & 2s(3)\beta(3) \\ 1s(4)\alpha(4) & 1s(4)\beta(4) & 2s(4)\alpha(4) & 2s(4)\beta(4) \end{vmatrix}$$

Computational problems

Before solving the computational problems, it is recommended that students work through Tutorials 1–3 under the Help menus in Spartan Student Edition to gain familiarity with the program.

Computational Problem 10.1: Calculate the total energy and $1s$ orbital energy for Ne using the Hartree–Fock method and the (a) 3-21G, (b) 6-31G*, and (c) 6-311+G** basis sets. Note the number of basis functions used in the calculations. Calculate the relative error of your result compared with the Hartree–Fock limit of -128.854705 hartree for each basis set. Rank the basis sets in terms of their approach to the Hartree–Fock limit for the total energy.

Step 1: Create a new file, and enter neon using the inorganic palette which is accessed by clicking the "Inorganic" tab on the model kit with a single valence on the atom. Delete the valence. Enter the various basis sets within the same file using the "New Molecule" command in the File menu.

Step 2: Go to Setup → Calculations. Set the calculation type to "Energy" and the method to Hartree–Fock with the appropriate basis set. "Total Charge" should be neutral. Enter the appropriate number of unpaired electrons. "Global Calculations" should be checked. Click "Submit." Your calculations will start immediately.

Step 3: Once your calculations complete, go to Display → Output. Note the values of the total energy for each calculation.

Step 4: Make a table showing the basis set, number of basis functions, total energy, and relative error with respect to the Hartree–Fock limit. Rank the basis sets in terms of their approach to the Hartree–Fock limit for the total energy.

Computational Problem 10.2: Calculate the total energy and $4s$ orbital energy for K using the Hartree–Fock method and the (a) 3-21G and (b) 6-31G* basis sets. Note the number of basis functions used in the calculations. Calculate the percentage deviation from the Hartree–Fock limits which are -16245.7 eV for the total energy and -3.996 eV for the $4s$ orbital energy. Rank the basis sets in terms of their approach to the Hartree–Fock limit for the total energy. What percentage error in the Hartree–Fock limit to the total energy corresponds to a typical reaction enthalpy change of 100 kJ mol^{-1}?

Step 1: Create a new file, and enter K using the inorganic palette which is accessed by clicking the "Inorganic" tab on the model kit with a single valence on the atom. Delete the valence. Enter the various basis sets within the same file using the "New Molecule" command in the File menu.

Step 2: Go to Setup → Calculations. Set the calculation type to "Energy" and the method to Hartree–Fock with the appropriate basis set. "Total Charge" should be neutral. Enter the appropriate number of unpaired electrons. "Global Calculations" should be checked. Click "Submit." Your calculations will start immediately.

Step 3: Once your calculations complete, go to Display → Output. Note the values of the total energy and 4s orbital energy for each calculation

Step 4: Make a table showing the basis set, total energy, relative error of the total energy with respect to the Hartree–Fock limit, orbital energy, relative error of the orbital energy with respect to the Hartree–Fock limit. Rank the basis sets in terms of their approach to the Hartree–Fock limit for the total energy.

Computational Problem 10.3: Calculate the ionization energy for (a) Li, (b) F, (c) S, (d) Cl, and (e) Ne using the Hartree–Fock method and the 6-311+G** basis set. Carry out the calculation in two different ways: (a) Use Koopmans' theorem and (b) compare the total energy of the neutral and singly ionized atom. Compare your answers with literature values.

Step 1: Create a new file, and enter Li using the inorganic palette which is accessed by clicking the "Inorganic" tab on the model kit with a single valence on the atom. Delete the valence. Hartree–Fock method and the 6-311+G** basis set. Enter the appropriate number of unpaired electrons. Uncheck the "global calculations" box. Using the "New Molecule" command in the File menu, enter the other atoms and the appropriate cation for each atom. Check the orbitals and energies box. The global calculations box should not be checked. Submit the calculation.

Step 2: Examine the output file and record the energy of the highest occupied AO, the total energy of the neutral atom, the total energy of the cation, the difference in energy of these two previous values, and literature values for the ionization energy as separate columns in a table. Discuss your results by comparing the literature and calculated results.

Computational Problem 10.4: Calculate the electron affinity for (a) Li, (b) F, (c) S, and (d) Cl using the Hartree–Fock method and the 6-311+G** basis set by comparing the total energy of the neutral and singly ionized atom. Compare your answers with literature values.

Step 1: Create a new file, and enter Li using the inorganic palette which is accessed by clicking the "Inorganic" tab on the model kit with a single valence on the atom. Delete the valence. Hartree–Fock method and the 6-311+G** basis set. Enter the appropriate number of unpaired electrons. The global calculations box should not be checked. Using the "New Molecule" command in the File menu, enter the other atoms and the appropriate anion for each atom. Check the orbitals and energies box. Submit the calculation.

Step 2: Examine the output file and record the energy of the highest occupied AO, the total energy of the neutral atom, the total energy of the anion, the difference in energy of these two previous values, and literature values for the ionization energy as separate columns in a table. Discuss your results by comparing the literature and calculated results.

Computational Problem 10.5: Using your results from C10.3 and C10.4, calculate the Mulliken electronegativity for (a) Li, (b) F, (c) S, (d) Cl. Compare your results with literature values.

Computational Problem 10.6: To assess the accuracy of the Hartree–Fock method for calculating energy changes in reactions, calculate the total energy change for the reaction $CH_3OH \rightarrow CH_3 + OH$ by calculating the difference in the total energy of reactants and products using the Hartree–Fock method and the 6-31G* basis set. Compare your result with a calculation using the B3LYP method and the same basis set and with the experimental value of 410. kJ mol^{-1}. As discussed in Chapter 4, the B3LYP method takes electron correlation into account. What percentage error in the Hartree–Fock total energy for CH_3OH would account for the difference between the calculated and experimental value of ΔU?

Chapter 11: Quantum States for Many-Electron Atoms and Atomic Spectroscopy

P11.1) The principal line in the emission spectrum of potassium is violet. On close examination, the line is seen to be a doublet with wavelengths of 393.366 and 396.847 nm. Explain the source of this doublet.

The lower lying state is the level and state $^2S_{1/2}$, and the upper level is ^2P, which contains the states $^2P_{1/2}$ and $^2P_{3/2}$. Both transitions are allowed, and the 589.0 nm wavelength corresponds to the transition between the ground state and $^2P_{3/2}$ because states with higher J lie lower in energy. The 589.6 nm wavelength corresponds to the transition between the ground state and $^2P_{1/2}$.

P11.5) What J values are possible for a ^6H term? Calculate the number of states associated with each level and show that the total number of states is the same as that calculated from $(2S + 1)(2L + 1)$.

$S = 5/2$, $L = 5$.

J lies between $|L + S|$ and $|L - S|$ and can have the values 15/2, 13/2, 11/2, 9/2, 7/2, and 5/2. The number of states is $2J + 1$ or 16, 14, 12, 10, 8, and 6, respectively.

This gives a total number of states of $16 + 14 + 12 + 10 + 8 + 6 = 66$.

$(2S + 1)(2L + 1) = 6 \times 11 = 66$ also.

P11.12) Calculate the wavelengths of the first three lines of the Lyman, Balmer, and Paschen series, and the series limit (the shortest wavelength) for each series.

Lyman Series: $E_n = R_H \left(\dfrac{1}{1^2} - \dfrac{1}{n^2} \right)$

$n = 2 \quad E_2 = R_H \left(1 - \dfrac{1}{4} \right) = \dfrac{3}{4} R_H = 82257.8 \text{ cm}^{-1} \quad \lambda = 121.569 \text{ nm}$

$n = 3 \quad E_2 = R_H \left(1 - \dfrac{1}{9} \right) = \dfrac{8}{9} R_H = 97490.7 \text{ cm}^{-1} \quad \lambda = 102.574 \text{ nm}$

$n = 4 \quad E_2 = R_H \left(1 - \dfrac{1}{16} \right) = \dfrac{15}{16} R_H = 102822 \text{ cm}^{-1} \quad \lambda = 97.2553 \text{ nm}$

$n = \infty \quad E_2 = R_H \left(1 - \dfrac{1}{\infty} \right) = R_H = 109678 \text{ cm}^{-1} \quad \lambda = 91.1768 \text{ nm}$

Balmer Series: $E_n = R_H \left(\dfrac{1}{2^2} - \dfrac{1}{n^2} \right)$

$n = 3 \quad E_2 = R_H \left(\dfrac{1}{4} - \dfrac{1}{9} \right) = \dfrac{5}{36} R_H = 15232.9 \text{ cm}^{-1} \quad \lambda = 656.473 \text{ nm}$

$n = 3 \quad E_2 = R_H \left(\dfrac{1}{4} - \dfrac{1}{16} \right) = \dfrac{3}{16} R_H = 20564.4 \text{ cm}^{-1} \quad \lambda = 486.276 \text{ nm}$

$n = 4 \quad E_2 = R_H \left(\dfrac{1}{4} - \dfrac{1}{25} \right) = \dfrac{21}{100} R_H = 23032.2 \text{ cm}^{-1} \quad \lambda = 434.175 \text{ nm}$

$n = \infty \quad E_2 = R_H \left(\dfrac{1}{4} - \dfrac{1}{\infty} \right) = \dfrac{1}{4} R_H = 27419.3 \text{ cm}^{-1} \quad \lambda = 364.707 \text{ nm}$

Paschen Series: $E_n = R_H \left(\dfrac{1}{3^2} - \dfrac{1}{n^2} \right)$

$n = 4 \quad E_2 = R_H \left(\dfrac{1}{9} - \dfrac{1}{16} \right) = \dfrac{7}{144} R_H = 5331.52 \text{ cm}^{-1} \quad \lambda = 1875.64 \text{ nm}$

$n = 5 \quad E_2 = R_H \left(\dfrac{1}{9} - \dfrac{1}{25} \right) = \dfrac{16}{225} R_H = 7799.25 \text{ cm}^{-1} \quad \lambda = 1282.17 \text{ nm}$

$n = 6 \quad E_2 = R_H \left(\dfrac{1}{9} - \dfrac{1}{36} \right) = \dfrac{1}{12} R_H = 9139.75 \text{ cm}^{-1} \quad \lambda = 1094.12 \text{ nm}$

$n = \infty \quad E_2 = R_H \left(\dfrac{1}{9} - \dfrac{1}{\infty} \right) = \dfrac{1}{9} R_H = 12186 \text{ cm}^{-1} \quad \lambda = 820.591 \text{ nm}$

P11.14) The inelastic mean free path of electrons in a solid, λ, governs the surface sensitivity of techniques such as AES and XPS. The electrons generated below the surface must make their way to the surface without losing energy in order to give elemental and chemical shift information. An empirical expression for elements that give λ as a function of the kinetic energy of the electron generated in AES or XPS is $\lambda = 538E^{-2} + 0.41(lE)^{0.5}$. The units of λ are monolayers, E is the kinetic energy of the electron, and l is the monolayer thickness in nanometers. On the basis of this equation, what kinetic energy maximizes the surface sensitivity for a monolayer thickness of 0.3 nm? An equation solver would be helpful in obtaining the answer.

$$\frac{d}{dE} \left(538E^{-2} + 0.41 (lE)^{0.5} \right) = -\frac{1076}{E^3} + \frac{0.1123}{\sqrt{E}}$$

Setting the derivative equal to zero and solving for E gives two complex roots and $E = 39$ eV.

P11.17) What are the levels that arise from a ^4F term? How many states are there in each level?

$S = 3/2$ and $L = 3$. M_L can range from $-L$ to $+L$ and can have the values $-3, -2, -1, 0, 1, 2$, and 3 in this case. M_S can range from $-S$ to $+S$ and can have the values $-3/2, -1/2, 1/2$, and $3/2$ in this case. J lies between $|L + S|$ and $|L - S|$ and can have the values $9/2, 7/2, 5/2$, and $3/2$ for this case. Because the number of states is $2J+1$, these levels have 10, 8, 6, and 4 states respectively.

P11.23) Use the transition frequencies shown in Example Problem 11.7 to calculate the energy (in joules and electron-volts) of the six levels relative to the $3s\ ^2S_{1/2}$ level. State your answers with the correct number of significant figures.

$$E\left(3p\,^2P_{1/2}\right) = \frac{hc}{\lambda} = \frac{6.626\times10^{-34}\ \text{J s}\times2.998\times10^8\ \text{m s}^{-1}}{589.6\times10^{-9}\ \text{m}} = 3.369\times10^{-19}\ \text{J} = 2.103\ \text{eV}$$

$$E\left(3p\,^2P_{3/2}\right) = \frac{hc}{\lambda} = \frac{6.626\times10^{-34}\ \text{J s}\times2.998\times10^8\ \text{m s}^{-1}}{589.0\times10^{-9}\ \text{m}} = 3.373\times10^{-19}\ \text{J} = 2.105\ \text{eV}$$

$$E\left(4s\,^2S_{1/2}\right) = \frac{hc}{\lambda} = \frac{6.626\times10^{-34}\ \text{J s}\times2.998\times10^8\ \text{m s}^{-1}}{589.6\times10^{-9}\ \text{m}} + \frac{6.626\times10^{-34}\ \text{J s}\times2.998\times10^8\ \text{m s}^{-1}}{1183.3\times10^{-9}\ \text{m}}$$

$$= 5.048\times10^{-19}\,\text{J} = 3.150\ \text{eV}$$

$$E\left(5s\,^2S_{1/2}\right) = \frac{hc}{\lambda} = \frac{6.626\times10^{-34}\ \text{J s}\times2.998\times10^8\ \text{m s}^{-1}}{589.0\times10^{-9}\ \text{m}} + \frac{6.626\times10^{-34}\ \text{J s}\times2.998\times10^8\,\text{m s}^{-1}}{616.0\times10^{-9}\ \text{m}}$$

$$= 6.597\times10^{-19}\,\text{J} = 4.118\ \text{eV}$$

$$E\left(3d\,^2D_{3/2}\right) = \frac{hc}{\lambda} = \frac{6.626\times10^{-34}\ \text{J s}\times2.998\times10^8\ \text{m s}^{-1}}{589.6\times10^{-9}\text{m}} + \frac{6.626\times10^{-34}\ \text{J s}\times2.998\times10^8\ \text{m s}^{-1}}{818.3\times10^{-9}\text{m}}$$

$$= 5.797\times10^{-19}\,\text{J} = 3.618\ \text{eV}$$

$$E\left(4d\,^2D_{3/2}\right) = \frac{hc}{\lambda} = \frac{6.626\times10^{-34}\ \text{J s}\times2.998\times10^8\ \text{m s}^{-1}}{589.0\times10^{-9}\ \text{m}} + \frac{6.626\times10^{-34}\ \text{J s}\times2.998\times10^8\ \text{m s}^{-1}}{568.2\times10^{-9}\ \text{m}}$$

$$= 6.869\times10^{-19}\,\text{J} = 4.287\ \text{eV}$$

P11.25) The spectrum of the hydrogen atom reflects the splitting of the $1s^2S$ and $2p^2P$ terms into levels. The energy difference between the levels in each term is much smaller than the difference in energy between the terms. Given this information, how many spectral lines are observed in the $1s^2S \rightarrow 2p^2P$ transition? Are the frequencies of these transitions very similar or quite different? The ^2S term has a single level, $^2S_{1/2}$. The ^2P term splits into two levels, $^2P_{1/2}$ and $^2P_{3/2}$. Therefore, there will be two closely spaced lines in the spectrum corresponding to the transitions $^2S_{1/2} \rightarrow {}^2P_{1/2}$ and

$^2S_{1/2} \rightarrow {}^2P_{3/2}$. The energy spacing between the lines will be much smaller than the energy of the transition.

P11.27) What atomic terms are possible for the following electron configurations? Which of the possible terms has the lowest energy?

a) ns^1np^1 b) ns^1nd^1 c) ns^2np^1 d) ns^1np^2

a) ns^1np^1 L can only have the value 1, and S can have the values 0 and 1. The possible terms are ^1P and ^3P. Hund's Rules predict that the ^3P term will have the lower energy.

b) ns^1nd^1 L can only have the value 2, and S can have the values 0 and 1. The possible terms are ^1D and ^3D. Hund's Rules predict that the ^3D term will have the lower energy.

c) ns^2np^1 L can only have the value 1, and S can only have the value 1/2. The only possible term is ^2P.

d) ns^1np^2 A table such as the table in the text for the p^2 configuration will have three columns, one for each of the electrons, for M_L and M_S. Each of the fifteen states for the p^2 configuration can be combined with $m_s = \pm\dfrac{1}{2}$ for the ns electron. This gives a total of 30 states. Working through the table gives ^2D, ^4P, ^2P, and ^2S terms. Hund's Rules predict that the ^4P term will have the lowest energy.

P11.28) Two angular momenta with quantum numbers $j_1 = 3/2$ and $j_2 = 5/2$ are added. What are the possible values of J for the resultant angular momentum states?

$J = |J_1 + J_2|, |J_1 + J_2 - 1|, |J_1 + J_2 - 2|, ..., |J_1 - J_2|$ giving possible J values of 4, 3, 2, and 1.

P11.29) Derive the ground-state term symbols for the following configurations:

a. d^2 b. f^9 c. f^{12}

The method illustrated in Example Problem 11.4 is used for all parts.

a)

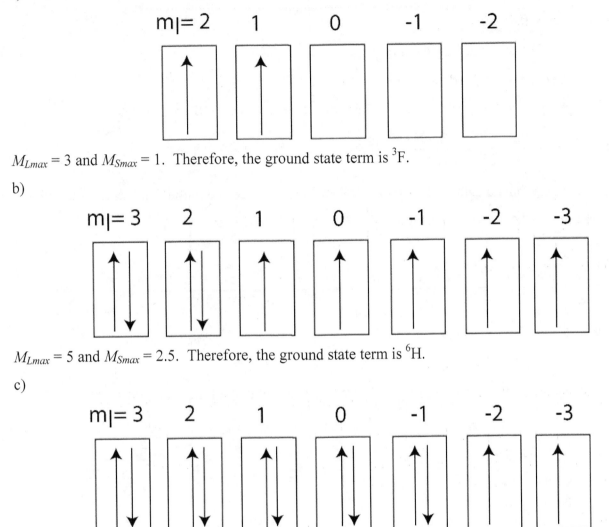

$M_{Lmax} = 3$ and $M_{Smax} = 1$. Therefore, the ground state term is ^3F.

b)

$M_{Lmax} = 5$ and $M_{Smax} = 2.5$. Therefore, the ground state term is ^6H.

c)

$M_{Lmax} = 5$ and $M_{Smax} = 1$. Therefore, the ground state term is ^3H.

P11.30) The first ionization potential of ground-state He is 24.6 eV. The wavelength of light associated with the $1s2p\ ^1P$ term is 58.44 nm. What is the ionization energy of the He atom in this excited state?

The photon energy is $E = \dfrac{hc}{\lambda} = \dfrac{6.626 \times 10^{-34}\ \text{J s} \times 2.998 \times 10^8\ \text{m s}^{-1}}{58.44 \times 10^{-9}\ \text{m}} \times \dfrac{1\ \text{eV}}{1.602 \times 10^{-19}\ \text{J}} = 21.2\ \text{eV}$

Therefore, the ionization energy of the He atom in this state is 24.6 eV − 21.2 eV = 3.4eV.

P11.31) In the Na absorption spectrum, the following transitions are observed:

$4p\,^2P \rightarrow 3s\,^2S$ $\lambda = 330.26$ nm

$3p\,^2P \rightarrow 3s\,^2S$ $\lambda = 589.593$ nm, 588.996 nm

$5s\,^2S \rightarrow 3p\,^2P$ $\lambda = 616.073$ nm, 615.421 nm

Calculate the energies of the $4p\,^2P$ and $5s\,^2S$ states with respect to the $3s\,^2S$ ground state.

$$E\left(4p\,^2P\right) = \frac{hc}{\lambda} = \frac{6.626\times10^{-34}\ \text{J s}\times2.998\times10^8\ \text{m s}^{-1}}{330.26\times10^{-9}\ \text{m}} = 6.015\times10^{-19}\ \text{J} = 3.754\ \text{eV}$$

By looking at the Grotrian diagram of Example Problem 11.2, it is seen that the $5s\,^2S$ state is accessed by absorption of the photons of wavelength 588.996 nm and 616.073 nm.

$$E\left(5s\,^2S\right) = \frac{hc}{\lambda} = \frac{6.626\times10^{-34}\ \text{J s}\times2.998\times10^8\ \text{m s}^{-1}}{588.996\times10^{-9}\ \text{m}} + \frac{6.626\times10^{-34}\ \text{J s}\times2.998\times10^8\ \text{m s}^{-1}}{616.073\times10^{-9}\ \text{m}}$$

$$= 6.597\times10^{-19}\ \text{J} = 4.117\ \text{eV}$$

P11.33) List the quantum numbers L and S that are consistent with the following terms:

a.　^4S　　b.　^4G　　c.　^3P　　d.　^2D

a) ^4S: $L = 0$, $2S + 1 = 4$, $S = 3/2$

b) ^4G: $L = 4$, $2S + 1 = 4$, $S = 3/2$

c) ^3P: $L = 1$, $2S + 1 = 3$, $S = 1$

d) ^2D: $L = 2$, $2S + 1 = 2$, $S = \frac{1}{2}$

P11.36) A general way to calculate the number of states that arise from a given configuration is as follows. Calculate the combinations of m_l and m_s for the first electron, and call that number n. The number of combinations used is the number of electrons, which we call m. The number of unused combinations is $n–m$. According to probability theory, the number of distinct permutations that arise from distributing the m electrons among the n combinations is $n!/[m!(n–m)!]$. For example, the number of states arising from a p^2 configuration is $6!/[2!4!]=15$, which is the result obtained in Section 11.2. Using this formula, calculate the number of possible ways to place five electrons in a d subshell. What is the ground-state term for the d^5 configuration and how many states does the term include?

The first electrons can have any combination of 5 m_l and 2 m_s values so that $n = 10$ and $m = 5$. Using

the formula, the calculated number of states is $\dfrac{10!}{5!(10-5)!} = 252$. The number of states in a term is

$(2L+1)(2S+1)$. The terms from Table 11.1 are listed here, with the number of states in each term in

square brackets following the term designation: $^6S[6 \times 1]$, $^4G[4 \times 9]$, $^4F[4 \times 7]$, $^4D[4 \times 5]$, $^4P[4 \times 3]$,

$^2I[2 \times 13]$, $^2H[2 \times 11]$, $^2G(2)[2 \times 2 \times 9]$, $^2F(2)[2 \times 2 \times 7]$, $^2D(3)[3 \times 2 \times 5]$, $^2P[2 \times 3]$, and $^2S[2 \times 1]$. The

number of states in these terms is $6 + 36 + 28 + 20 + 12 + 26 + 22 + 36 + 28 + 30 + 6 + 2 = 252$,

showing consistency.

The ground state is that with the maximum multiplicity which according to Table 10.5 is the ^6S term.

The number of states in the term $= (2L + 1)(2S + 1) = 6$.

Chapter 12: The Chemical Bond in Diatomic Molecules

P12.2) The overlap integral for ψ_g and ψ_u as defined in Section 12.3 is given by

$$S_{ab} = e^{-\zeta R/a_0} \left(1 + \zeta \frac{R}{a_0} + \frac{1}{3}\zeta^2 \frac{R^2}{a_0^2} \right)$$

Plot S_{ab} as a function of R/a_0 for ζ =0.8, 1.0, and 1.2. Estimate the value of R/a_0 for which S_{ab}=0.4 for each of these values of ζ.

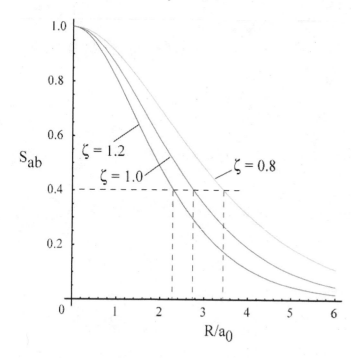

S_{ab} has the value of 0.4 at R/a_0 = 3.42, 2.68, and 2.25 for ζ = 0.8, 1.0, and 1.2, respectively.

P12.3) Sketch out a molecular orbital energy diagram for CO and place the electrons in the levels appropriate for the ground state. The AO ionization energies are O2s: 32.3 eV; O2p: 15.8 eV; C2s: 19.4 eV; and C2p: 10.9 eV. The MO energies follow the sequence (from lowest to highest) $1\sigma, 2\sigma, 3\sigma, 4\sigma, 1\pi, 5\sigma, 2\pi, 6\sigma$. Assume that the 1σ and 2σ MOs originate from the $1s$ AOs and that the

3σ and 5σ MOs originate from the $2s$ AOs on C and O. Connect each MO level with the level of the major contributing AO on each atom.

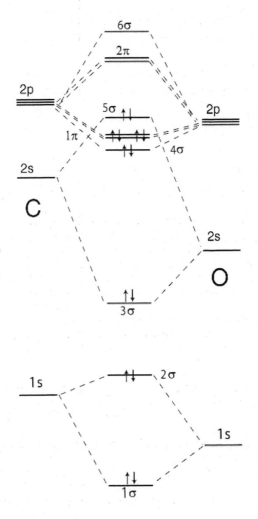

P12.9) Calculate the value for the coefficients of the AOs in Example Problem 12.3 for $S_{12} = 0.6$. How are they different from the values calculated in that problem for $S_{12} = 0.3$? Can you offer an explanation for the changes?

We first obtain the result $H_{12} = -1.75 \, S_{12} \sqrt{H_{11} H_{22}} = -16.7$ eV

Solving for the energies of ε_1 and ε_2 (bonding and antibonding MOs, respectively) gives the values $\varepsilon_1 = -20.9$ eV, and $\varepsilon_2 = +1.94$ eV.

We calculate $\dfrac{c_{12}}{c_{22}}$ by substituting the values for ε_1 and ε_2 in the first of the

Equations (12.9). Both equations give the same result.

$$c_{12}(H_{11} - \varepsilon) + c_{22}(H_{12} - \varepsilon S_{12}) = 0.$$

For $\varepsilon_2 = 1.94$ eV, $c_{12}(-13.6 - 1.94) + c_{22}(-16.7 - 0.6 \times 1.94) = 0$

$$\frac{c_{12}}{c_{22}} = -1.15$$

Using this result in the normalization equation $c_{12}^2 + c_{22}^2 + 2c_{12}c_{22}S_{12} = 1.$

$$(-1.15c_{22})^2 + c_{22}^2 + 2(-1.15c_{22}) \times 0.6 = 1$$

$$1.3225c_{22}^2 + c_{22}^2 - 1.38c_{22} = 0; \quad c_{22} = -1.03$$

$$c_{12} = 1.18, \quad \text{and } \sigma_2 = 1.18\phi_{H1s} - 1.03\phi_{F2p_z}$$

$$c_{12}(H_{11} - \varepsilon) + c_{22}(H_{12} - \varepsilon S_{12}) = 0.$$

For $\varepsilon_2 = -20.9$ eV, $c_{11}(-13.6 + 20.9) + c_{21}(-16.7 + 0.6 \times 20.9) = 0$

$$\frac{c_{11}}{c_{21}} = 0.57$$

Using this result in the normalization equation $c_{11}^2 + c_{21}^2 + 2c_{11}c_{21}S_{12} = 1.$

$$(0.57c_{21})^2 + c_{21}^2 + 2(0.57c_{21}^2) \times 0.6 = 1$$

$$0.3249c_{21}^2 + c_{21}^2 + 0.684c_{21}^2 = 0; \quad c_{21} = 0.71$$

$$c_{11} = 0.40, \quad \text{and } \sigma_2 = 0.40\phi_{H1s} + 0.71\phi_{F2p_z}$$

The increase in the overlap results in $c_{11} = 0.40$ and $c_{21} = 0.71$ for $\varepsilon_1 = -20.9$ eV and in

$c_{22} = -1.03$ and $c_{12} = 1.18$ for $\varepsilon_2 = +1.94$ eV. As for $S_{12} = 0.3$, the coefficient on the lower lying AO is

greater for the bonding orbital and less for the antibonding orbital. Also as before, the signs of the

coefficients are the same for the bonding orbital and opposite for the antibonding orbital. However, the

magnitude of the coefficients is more nearly equal, due to the greater interaction that arises from a

greater overlap, because H_{12} increases linearly with S_{12}. In the bonding MO, the electron is shared more

equally by the two atoms for the greater overlap.

P12.10 Arrange the following in terms of decreasing bond energy and bond length:

O_2^+, O_2, O_2^-, and O_2^{2-}.

$O_2^{2-}: \left(1\sigma_g\right)^2 \left(1\sigma_u^*\right)^2 \left(2\sigma_g\right)^2 \left(2\sigma_u^*\right)^2 \left(3\sigma_g\right)^2 \left(1\pi_u\right)^2 \left(1\pi_u\right)^2 \left(1\pi_g^*\right)^2 \left(1\pi_g^*\right)^2$

Bond Order $= \dfrac{10-8}{2} = 1$

$O_2^-: \left(1\sigma_g\right)^2 \left(1\sigma_u^*\right)^2 \left(2\sigma_g\right)^2 \left(2\sigma_u^*\right)^2 \left(3\sigma_g\right)^2 \left(1\pi_u\right)^2 \left(1\pi_u\right)^2 \left(1\pi_g^*\right)^2 \left(1\pi_g^*\right)^1$

Bond Order $= \dfrac{10-7}{2} = 1.5$

$O_2^-: \left(1\sigma_g\right)^2 \left(1\sigma_u^*\right)^2 \left(2\sigma_g\right)^2 \left(2\sigma_u^*\right)^2 \left(3\sigma_g\right)^2 \left(1\pi_u\right)^2 \left(1\pi_u\right)^2 \left(1\pi_g^*\right)^1 \left(1\pi_g^*\right)^1$

Bond Order $= \dfrac{10-6}{2} = 2$

$O_2^+: \left(1\sigma_g\right)^2 \left(1\sigma_u^*\right)^2 \left(2\sigma_g\right)^2 \left(2\sigma_u^*\right)^2 \left(3\sigma_g\right)^2 \left(1\pi_u\right)^2 \left(1\pi_u\right)^2 \left(1\pi_g^*\right)^1$

Bond Order $= \dfrac{10-5}{2} = 2.5$

Bond energy: $O_2^+ > O_2 > O_2^- > O_2^{2-}$

Bond length: $O_2^{-2} > O_2^- > O_2 > O_2^+$

P12.13) What is the electron configuration corresponding to O_2, O_2^-, and O_2^+? What do you expect the relative order of bond strength to be for these species? Which, if any, have unpaired electrons?

$O_2: \left(1\sigma_g\right)^2 \left(1\sigma_u^*\right)^2 \left(2\sigma_g\right)^2 \left(2\sigma_u^*\right)^2 \left(3\sigma_g\right)^2 \left(1\pi_u\right)^2 \left(1\pi_u\right)^2 \left(1\pi_g^*\right)\left(1\pi_g^*\right)$

Bond Order $= \dfrac{10-6}{2} = 2$

$O_2^-: \left(1\sigma_g\right)^2 \left(1\sigma_u^*\right)^2 \left(2\sigma_g\right)^2 \left(2\sigma_u^*\right)^2 \left(3\sigma_g\right)^2 \left(1\pi_u\right)^2 \left(1\pi_u\right)^2 \left(1\pi_g^*\right)^2 \left(1\pi_g^*\right)^1$

Bond Order $= \dfrac{10-7}{2} = 1.5$

$O_2^+: \left(1\sigma_g\right)^2 \left(1\sigma_u^*\right)^2 \left(2\sigma_g\right)^2 \left(2\sigma_u^*\right)^2 \left(3\sigma_g\right)^2 \left(1\pi_u\right)^2 \left(1\pi_u\right)^2 \left(1\pi_g^*\right)^1$

Bond Order $= \dfrac{10-5}{2} = 2.5$

Because the bond strengths increase with increasing bond order, the relative order of bond strength is

$O_2^+ > O_2 > O_2^-$. O_2 has two unpaired electrons, and the positively and negatively charged species each

have one.

P12.15) Evaluate the energy for the two MOs generated by combining two H1s AOs. Use Equation (12.23) and carry out the calculation for $S_{12} = 0.1, 0.2,$ and 0.6 to mimic the effect of decreasing the atomic separation in the molecule. Use the parameters $H_{11} = H_{22} = -13.6$ eV and $H_{12} = -1.75 S_{12} \sqrt{H_{11} H_{22}}$. Explain the trend that you observe in the results.

$$E_1 = \frac{H_{11} + H_{12}}{1 + S_{12}} \text{ and } E_2 = \frac{H_{11} - H_{12}}{1 - S_{12}}$$

S_{12}	H_{12}(eV)	ε_1(eV)	ε_2(eV)
0.1	−2.38	−14.5	−12.5
0.2	−4.76	−15.3	−11.1
0.6	−14.3	−17.4	+1.7

The increase in the overlap mimics a decrease in the bond length. As S_{12} increases, the orbitals overlap more and their interaction becomes greater. The result is that ε_1 becomes more strongly binding and ε_2 becomes more strongly antibonding.

P12.16) Show that calculating E_u in the manner described by Equation (12.7) gives the result $E_u = (H_{aa} - H_{ab})/(1 - S_{ab})$.

$$E_u = \frac{\int \psi_u^* \hat{H} \psi_u \, d\tau}{\int \psi_u^* \psi_u \, d\tau}$$

$$= \frac{1}{2(1-S_{ab})} \left(\int \psi_{H1s_a}^* \hat{H} \psi_{H1s_a} \, d\tau + \int \psi_{H1s_b}^* \hat{H} \psi_{H1s_b} \, d\tau - \int \psi_{H1s_b}^* \hat{H} \psi_{H1s_a} \, d\tau - \int \psi_{H1s_a}^* \hat{H} \psi_{H1s_b} \, d\tau \right)$$

$$= \frac{1}{2(1-S_{ab})} \left(H_{aa} + H_{bb} - H_{ba} - H_{ab} \right)$$

$$= \frac{H_{aa} - H_{ab}}{1 + S_{ab}}$$

P12.17) A surface displaying a contour of the total charge density in LiH is shown here. What is the relationship between this surface and the MOs displayed in Problem P12.12? Why does this surface closely resemble one of the MOs?

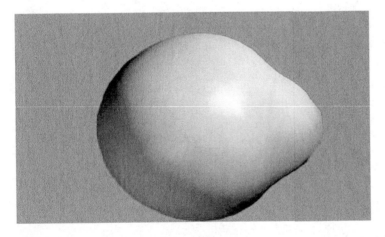

The surface of electron density will be determined by all filled MOs, but the unfilled MOs will not contribute. Therefore, the $3\sigma*$ MO has no influence in determining the electron density surface. Higher energy MOs are further out from the molecular skeleton than lower lying MOs. Therefore, they have more influence in determining the electron density contour. Therefore, the contour looks like the 2σ rather than the 1σ MO.

P12.23) Calculate the bond order in each of the following species. Which of the species in part a–d do you expect to have the shorter bond length?

a. Li_2 or Li_2^+

c. O_2 or O_2^+

b. C_2 or C_2^+

d. F_2 or F_2^-

Li_2: $\left(1\sigma_g\right)^2 \left(1\sigma_u^*\right)^2 \left(2\sigma_g\right)^2$

Bond Order $= \dfrac{4-2}{2} = 1$

Li_2^+: $\left(1\sigma_g\right)^2 \left(1\sigma_u^*\right)^2 \left(2\sigma_g\right)^1$

Bond Order $= \dfrac{3-2}{2} = 0.5$

C_2: $\left(1\sigma_g\right)^2 \left(1\sigma_u^*\right)^2 \left(2\sigma_g\right)^2 \left(2\sigma_u^*\right)^2 \left(1\pi_u\right)^4$

Bond Order $= \dfrac{8-4}{2} = 2$

C_2^+: $\left(1\sigma_g\right)^2 \left(1\sigma_u^*\right)^2 \left(2\sigma_g\right)^2 \left(2\sigma_u^*\right)^2 \left(1\pi_u\right)^3$

Bond Order $= \dfrac{7-4}{2} = 1.5$

O_2: $\left(1\sigma_g\right)^2 \left(1\sigma_u^*\right)^2 \left(2\sigma_g\right)^2 \left(2\sigma_u^*\right)^2 \left(3\sigma_g\right)^2 \left(1\pi_u\right)^2 \left(1\pi_u\right)^2 \left(1\pi_g^*\right)^1 \left(1\pi_g^*\right)^1$

Bond Order $= \dfrac{10-6}{2} = 2$

O_2^+: $\left(1\sigma_g\right)^2 \left(1\sigma_u^*\right)^2 \left(2\sigma_g\right)^2 \left(2\sigma_u^*\right)^2 \left(3\sigma_g\right)^2 \left(1\pi_u\right)^2 \left(1\pi_u\right)^2 \left(1\pi_g^*\right)^1$

Bond Order $= \dfrac{10-5}{2} = 2.5$

F_2: $\left(1\sigma_g\right)^2 \left(1\sigma_u^*\right)^2 \left(2\sigma_g\right)^2 \left(2\sigma_u^*\right)^2 \left(3\sigma_g\right)^2 \left(1\pi_u\right)^2 \left(1\pi_u\right)^2 \left(1\pi_u^*\right)^2 \left(1\pi_g^*\right)^2$

Bond Order $= \dfrac{10-8}{2} = 1$

F_2^-: $\left(1\sigma_g\right)^2 \left(1\sigma_u^*\right)^2 \left(2\sigma_g\right)^2 \left(2\sigma_u^*\right)^2 \left(3\sigma_g\right)^2 \left(1\pi_u\right)^2 \left(1\pi_u\right)^2 \left(1\pi_g^*\right)^3 \left(1\pi_g^*\right)^2$

Bond Order $= \dfrac{10-9}{2} = 0.5$

Because the bond length is shorter for a greater bond order, the answers are Li_2, C_2, O_2^+, F_2.

Computational problems

Before solving the computational problems, it is recommended that students work through Tutorials 1–3 under the Help menus in Spartan Student Edition to gain familiarity with the program.

Computational Problem 12.1:

According to Hund's rules, the ground state of O_2 should be a triplet because the last two electrons are placed in a doubly degenerate set of π MOs. Calculate the energy of the singlet and triplet states of O_2 using the B3LYP method and the 6-31G* basis set. Does the singlet or triplet have the lower energy? Both states will be populated if the energy difference $\Delta E \sim kT$. For which temperature is this the case?

Step 1: Create a new file, and choose O= from the inorganic palette which is accessed by clicking the "Inorganic" tab on the model kit. Add a second O= to the initial atom.

Step 2: Using the calculations menu, enter equilibrium geometry, the B3LYP method, the 6-31G* basis set and the appropriate number of unpaired electrons for the singlet state. Make sure that you check the Infrared spectra box. The global calculations box should not be checked.

Step 3: Repeat step 2 for the triplet state using the mew molecule command. Submit the calculation.

Step 4: Once your calculations complete, go to Display → Output. Note the values of the total energy for the triplet and singlet states. **You cannot conclude that the optimized structure corresponds to an equilibrium structure unless all the vibrational frequencies are real.** Go to display spectra under "Display" to see the vibrational frequencies.

Step 5: Calculate the temperature for which $\Delta E = kT$.

Computational Problem 12.2:

If the ground state of oxygen is a diradical, you might think that O_2 would dimerize to form square planar O_4 to achieve a molecule in which all electrons are paired. Optimize the geometry and calculate the energies of triplet O_2 and singlet O_4 using the B3LYP method and the 6-31G* basis set. Do you predict O_4 to be more or less stable than 2 O_2 molecules? Use a nonplanar shape in building your O_4 molecule.

Step 1: In this problem, it is important to start with a nonplanar geometry to avoid biasing the outcome of the calculation. Following these directions will create a "twisted square" initial geometry. Create a new file, and choose **nonplanar** trivalent O from the inorganic palette which is accessed by clicking the "Inorganic" tab on the model kit. Add a second and a third **nonplanar** trivalent O to the initial atom.

Use the "measure angle" button to determine the O-O-O angle. The result displayed in the bottom right corner of the window will be 109.47°. (Don't worry about the added H atoms. They will be removed in the next step.) Replace 109.47° with 95 and press enter. The O-O-O angle is now 95°.

Step 2: Hit the delete button and the H atoms will be removed. Add a fourth **nonplanar** trivalent O to the structure. Change the newly created O-O-O bond angle to 85° as in step 1. Using the dihedral angle button, measure the dihedral angle of the four O atoms. Change this value to 10°. Again use the delete button to remove the H atoms. Use the "make bond" button to complete the ring by clicking on the two closest valences. Use the delete button to remove the four dangling valences. You now have a "twisted square" starting geometry. Do not minimize the energy before continuing the calculation because the minimization will lead to a planar structure.

Step 3: Using the calculations menu, enter equilibrium geometry, the B3LYP method, the 6-31G* basis set and the appropriate number of unpaired electrons. Make sure that you check the Infrared spectra box. The global calculations box should not be checked. Submit the calculation.

Step 4: Once your calculations complete, go to Display → Output. Note the values of the total energy. **You cannot conclude that the optimized structure corresponds to an equilibrium structure unless all the vibrational frequencies are real.** Go to display spectra under "Display" to see the vibrational frequencies. If there is an equilibrium structure for O_4, describe it. Is the geometry optimized molecule planar or nonplanar?

Step 5: Compare the energy of O_4 to 2 molecules of O_2.

Computational Problem 12.3:

O_6 might be more stable than O_4 because the bond angle is larger, leading to less steric strain. Optimize the geometry and compare the energy of O_6 with 1.5 times the energy of O_4 using the B3LYP method and the 6-31G* basis set. Is O_6 more stable than O_4? Use a nonplanar shape in building your O_6 molecule. Is the geometry optimized molecule planar or nonplanar?

Step 1: In this problem, it is important to start with a nonplanar geometry to avoid biasing the outcome of the calculation. Following these directions will create a "boat" initial geometry. Create a new file, and choose **nonplanar** trivalent O from the inorganic palette which is accessed by clicking the "Inorganic" tab on the model kit. Add a second and then 3, 4, and 5 **nonplanar** trivalent O to the initial atom. Add to the appropriate valence to generate the "boat" structure.

Step 2: Use the "make bond" button to complete the ring by clicking on the two closest valences. Use the delete button to remove the six dangling valences. You now have a "boat" starting geometry. Minimize the energy.

Step 3: Using the calculations menu, enter equilibrium geometry, the B3LYP method, the 6-31G* basis set and the appropriate number of unpaired electrons. Make sure that you check the Infrared spectra box. The global calculations box should not be checked. Submit the calculation.

Step 4: Once your calculations complete, go to Display → Output. Note the values of the total energy. **You cannot conclude that the optimized structure corresponds to an equilibrium structure unless all the vibrational frequencies are real.** Go to display spectra under "Display" to see the vibrational frequencies. If there is an equilibrium structure for O_6, describe it. Is the geometry optimized molecule planar or nonplanar?

Step 5: Compare the energy of O_6 to 3 molecules of O_2. Compare the energy of O_6 to 1.5 times the energy of O_4. Is O_6 more stable than O_2? Is O_6 more stable than O_4?

Computational Problem 12.4:

In a LiF crystal, both the Li and F are singly ionized species. Optimize the geometry and calculate the charge on Li and F in a single LiFl molecule using the B3LYP method and the 6-31G* basis set. Are the atoms singly ionized? Compare the value of the bond length with the distance between Li^+ and F^- ions in the crystalline solid.

Step 1: Create a new file, and build the LiF molecule using the inorganic palette which is accessed by clicking the "Inorganic" tab on the model kit.

Step 2: Using the calculations menu, enter equilibrium geometry, the B3LYP method, the 6-31G* basis set and the appropriate number of unpaired electrons. Submit the calculation.

Step 3: Once your calculations complete, go to Display → Output. Note the values of the positions of Li and F and the charges on each atom. Are the atoms fully ionized? Is the LiF spacing equal to that in an LiF crystal?

Computational Problem 12.5:

Does LiF dissociate into neutral atoms or into Li^+ and F^-? Answer this question by comparing the energy difference between reactants and products for the reactions

$LiF(g) \rightarrow Li(g) + F(g)$ and $LiF(g) \rightarrow Li^+(g) + F^-(g)$ using the B3LYP method and the 6-31G* basis set.

Step 1: Create a new file, and build the LiF molecule using the inorganic palette which is accessed by clicking the "Inorganic" tab on the model kit. Using the "new molecule" command, add Li, F, Li^+, and F^- sequentially.

Step 2: Using the calculations menu, enter equilibrium geometry, the B3LYP method, the 6-31G* basis set and the appropriate number of unpaired electrons. Make sure the global calculations box is not checked. Submit the calculation.

Step 3: Once your calculations complete, go to Display → Output. Note the values of the energies of each species? Calculate the energy change for the reactions

$LiF(g) \rightarrow Li(g) + F(g)$ and $LiF(g) \rightarrow Li^+(g) + F^-(g)$. Decide whether dissociation into atoms or ions will occur.

Computational Problem 12.6:

Calculate Hartree–Fock MO energy values for HF using the MP2 method and the 6-31G* basis set. Make a molecular energy diagram to scale omitting the lowest energy MO. Why can you neglect this MO? Characterize the other MOs as bonding, antibonding, or nonbonding.

Step 1: Create a new file, and build the HF molecule using the inorganic palette which is accessed by clicking the "Inorganic" tab on the model kit.

Step 2: Using the calculations menu, enter equilibrium geometry, the MP2 method, the 6-31G* basis set and the appropriate number of unpaired electrons. Check the "orbitals and energies" box. Submit the calculation.

Step 3: Once your calculations complete, go to Display → Surfaces. By using the "add" button, add the LUMO, HOMO, and HOMO-x surfaces where x takes on integer values. What is the maximum value of x that you should enter to obtain surfaces for all occupied MOs?

Step 4: Look at each MO by checking the appropriate box. By clicking on the MO and going to the properties menu, you can make the MO transparent to better see the atom positions. Classify each MO as bonding, antibonding, and nonbonding. Explain how you reached your conclusions.

Computational Problem 12.7:

a) Based on the molecular orbital energy diagram in Problem C12.6, would you expect triplet neutral HF in which an electron is promoted from the 1π to the $4\sigma^*$ MO to be more or less stable than singlet HF? Because an electron is promoted from a nonbonding to an antibonding MO, the triplet is less stable.

b) Calculate the equilibrium bond length and total energy for singlet and triplet HF using the MP2 method and the 6-31G* basis set. Using the frequency as a criterion, are both stable molecules? Compare the bond lengths and vibrational frequencies.

Singlet: 0.934A, 4036 cm^{-1}, no imaginary frequencies, so molecule is stable

Triplet: Minimum energy not reached. Molecule is unstable and will dissociate into atoms.

c) Calculate the bond energy of singlet and triplet HF by comparing the total energies of the molecules with the total energy of F and H. Are your results consistent with the bond lengths and vibrational frequencies obtained in part b)?

Computational Problem 12.8:

Computational chemistry allows you to carry out calculations for hypothetical molecules that do not exist in order to see trends in molecular properties. Calculate the charge on the atoms in singlet HF and in triplet HF for which the bond length is fixed at 10% greater than the bond length for singlet HF. Are the trends that you see consistent with those predicted by Figure 12.4? Explain your answer.

Step 1: Create a new file, and build the HF molecule using the inorganic palette which is accessed by clicking the "Inorganic" tab on the model kit.

Step 2: Using the calculations menu, enter equilibrium geometry, the MP2 method, the 6-311+G** basis set and the appropriate number of unpaired electrons for the ground state of HF. Check the "orbitals and energies" box. The global calculations box should not be checked. Click "OK."

Step 3: Build another HF molecule in the same file using the "new molecule" command. For this triplet state molecule, you will constrain the H-F distance. Under the "Geometry menu," select "Constrain distance." Click on the "measure bond length" and on H and af using the shift key. Click on the open pink lock in the lower right corner of the window and change the H-F distance from the equilibrium value for singlet LiF of 0.934 A to a value that is 10% larger. Using the calculations menu, enter equilibrium geometry, the MP2 method, the 6-311+G** basis set and the appropriate number of unpaired electrons for the triplet state. Check the "orbitals and energies" box. Click "OK."

Step 4: Repeat step 3 for H-F distances that are 20 and 30% larger than 0.934A.

Step 5: Using the "New Molecule" command and the same basis set and method, set up calculations for singlet HF and for the total energy of the F atom.

Step 6: Submit your calculation.

Step 7: Make a table for the 3 H-F distances in which the columns are the total energy for the molecule, ΔE for the reaction HF $\rightarrow$ H + F, and the charge on the F atom. The rows list singlet HF and triplet HF for each of the H-F distances.

Step 8: Review Section 12.5 and discuss your results for the difference in the charge on the F atom for the singlet and triplet state. Explain the trend that you observe in the charge on the F atom with the H-F distance.

Chapter 13: Molecular Structure and Energy Levels for Polyatomic Molecules

P13.2) Predict whether LiH_2^+ and NH_2^- should be linear or bent based on the Walsh correlation diagram in Figure 13.10. Explain your answers.

The LiH_2^+ molecular ion has two valence electrons. These fill only the $1a_1$ MO. The correlation diagram shows that the energy of this MO is lowered if the molecule is bent. The molecule NH_2^- has eight valence electrons, as does H_2O, and is bent for the same reason.

P13.5) Use the method described in Example Problem 13.3 to show that the *sp*-hybrid orbitals

$\psi_a = 1/\sqrt{2}\,(-\phi_{2s} + \phi_{2p_z})$ and $\psi_b = 1/\sqrt{2}\,(-\phi_{2s} - \phi_{2p_z})$ are oriented 180° apart.

We differentiate ψ_a with respect to θ and set the derivative equal to zero.

$$\frac{d\psi_a}{d\theta} = \frac{1}{\sqrt{2}}\frac{d}{d\theta}\left[-\frac{1}{\sqrt{32\pi}}\left(\frac{\zeta}{a_0}\right)^{3/2}\left(2-\frac{r}{a_0}\right)e^{-r/a_0} + \frac{1}{\sqrt{32\pi}}\left(\frac{\zeta}{a_0}\right)^{3/2}\frac{r}{a_0}e^{-\zeta r/a_0}\cos\theta\right]$$

$$= -\frac{1}{\sqrt{32\pi}}\left(\frac{\zeta}{a_0}\right)^{3/2}\frac{r}{a_0}e^{-\zeta r/a_0}\sin\theta = 0$$

This equation is satisfied for $\theta = 0$ and $\theta = 180°$. The second derivative is used to establish which of these corresponds to a maximum.

$$\frac{d^2\psi_a}{d\theta^2} = -\frac{1}{\sqrt{32\pi}}\left(\frac{\varsigma}{a_0}\right)^{3/2}\left(\frac{r}{a_0}\right)e^{-\varsigma r/a_0}\cos\theta$$

$$= -\frac{1}{\sqrt{32\pi}}\left(\frac{\varsigma}{a_0}\right)^{3/2}\left(\frac{r}{a_0}\right)e^{-\varsigma r/a_0} \text{ at } \theta = 0$$

$$= \frac{1}{\sqrt{32\pi}}\left(\frac{\varsigma}{a_0}\right)^{3/2}\left(\frac{r}{a_0}\right)e^{-\varsigma r/a_0} \text{ at } \theta = 180°$$

At a maximum, $\dfrac{d^2\psi_a}{d\theta^2} < 0$ and at a minimum, $\dfrac{d^2\psi_a}{d\theta^2} > 0$. Therefore, the maximum is at 0°, and a

minimum at 180°. Applying the same procedure to ψ_b shows that the maximum is at 0°. Therefore, ψ_a

and ψ_b point in opposite directions separated by 180°.

P13.11) Determine the AO coefficients for the lowest energy Hückel π MO for butadiene.

The energy is given by $\varepsilon = \alpha + 1.62\beta$ and $\psi_\pi = c_1\psi_{2p_z}^a + c_2\psi_{2p_z}^b + c_3\psi_{2p_z}^c + \psi_{2p_z}^d$

Thus, the secular equations are

1): $c_1(H_{aa} - \varepsilon S_{aa}) + c_2(H_{ab} - \varepsilon S_{ab}) + c_3(H_{ac} - \varepsilon S_{ac}) + c_4(H_{ad} - \varepsilon S_{ad}) = 0$

2): $c_1(H_{ba} - \varepsilon S_{ba}) + c_2(H_{bb} - \varepsilon S_{bb}) + c_3(H_{bc} - \varepsilon S_{bc}) + c_4(H_{bd} - \varepsilon S_{bd}) = 0$

3): $c_1(H_{ca} - \varepsilon S_{ca}) + c_2(H_{cb} - \varepsilon S_{cb}) + c_3(H_{cc} - \varepsilon S_{cc}) + c_4(H_{cd} - \varepsilon S_{cd}) = 0$

4): $c_1(H_{da} - \varepsilon S_{da}) + c_2(H_{db} - \varepsilon S_{db}) + c_3(H_{dc} - \varepsilon S_{dc}) + c_4(H_{dd} - \varepsilon S_{dd}) = 0$

The overlap is given by $S_{jk} = \delta_{jk}$, and we assign values to the H_{jk} as follows:

$$H_{jk} = \begin{cases} \alpha \text{ if } j = k \\ \beta \text{ if } j \text{ and } k \text{ differ by 1, and} \\ 0 \text{ otherwise} \end{cases}$$

Substituting in the four secular equations gives the following relations:

1) $c_1\left[\alpha-(\alpha+1.62\beta)\right]+c_2\left[\beta\right]=0$

   We conclude that $c_2 = 1.62\,c_1$

2) $c_1\beta+c_2\left[\alpha-(\alpha+1.62\beta)\right]+c_3\beta=0$

   We conclude that $c_1 + c_3 = 1.62\,c_2$ or $c_3 = \left(1.62^2-1\right)c_1$

3) $c_2\beta+c_3\left[\alpha-(\alpha+1.62\beta)\right]+c_4\beta=0$

   We conclude that $c_2 + c_4 = 1.62\,c_3$ or $c_4 = \dfrac{\left(1.62^2-1\right)}{1.62}\,c_1$

The 4th equation needed to solve this system is the normalization condition, defined by

$$1 = \int\left(\psi_\pi\right)^*\psi_\pi\,d\tau = \int\left(c_1\psi_{2p_z}^a + c_2\psi_{2p_z}^b + c_3\psi_{2p_z}^c + c_4\psi_{2p_z}^d\right)^2 d\tau$$

$$= c_1^2\int\left(\psi_{2p_z}^a\right)^2 d\tau + c_2^2\int\left(\psi_{2p_z}^b\right)^2 d\tau + c_3^2\int\left(\psi_{2p_z}^c\right)^2 d\tau + c_4^2\int\left(\psi_{2p_z}^d\right)^2 d\tau$$

$$+\,2c_1c_2\int\psi_{2p_z}^a\psi_{2p_z}^b\,d\tau + 2c_1c_3\int\psi_{2p_z}^a\psi_{2p_z}^c\,d\tau + 2c_1c_4\int\psi_{2p_z}^a\psi_{2p_z}^d\,d\tau$$

$$+\,2c_2c_3\int\psi_{2p_z}^b\psi_{2p_z}^c\,d\tau + 2c_2c_4\int\psi_{2p_z}^b\psi_{2p_z}^d\,d\tau$$

$$= c_1^2 + c_2^2 + c_3^2 + c_4^2 + 2c_1c_2 S_{ab} + 2c_1c_3 S_{ac} + 2c_1c_4 S_{ad} + 2c_2c_3 S_{bc} + 2c_2c_4 S_{bd}$$

$$= c_1^2 + c_2^2 + c_3^2 + c_4^2$$

Substituting 1–3 into the normalization equation yields

$$c_1^2 + 1.62^2\,c_1^2 + \left(1.62^2-1\right)^2 c_1^2 + \left[\frac{\left(1.62^2-1\right)}{1.62}\right]^2 c_1^2 = 1$$

or

$$c_1^2 = \frac{1}{1+1.62^2+\left(1.62^2-1\right)^2+\left[\dfrac{1.62^2-1}{1.62}\right]^2}$$

and

$$c_1 = 0.3715$$
$$c_2 = 1.62 \cdot 0.3715 = 0.602$$
$$c_3 = \left(1.62^2 - 1\right)0.3715 = 0.602$$
$$c_4 = \left(1.62 - \frac{1}{1.62}\right)0.3715 = 0.3715$$
$$\psi_\pi = 0.3715\psi_{2p_z}^a + 0.602\psi_{2p_z}^b + 0.602\psi_{2p_z}^c + 0.3715\psi_{2p_z}^d$$

P13.13) Use the geometrical construction shown in Example Problem 13.10 to derive the π electron MO levels for the cyclopentadienyl radical. What is the total π energy of the molecule? How many unpaired electrons will the molecule have?

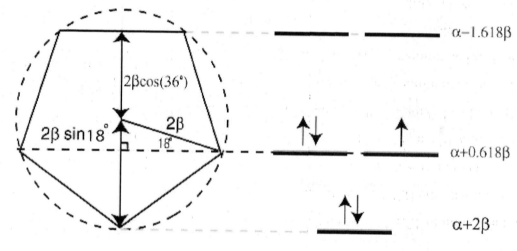

$$2\beta\sin\left(18°\right) = 0.618$$
$$2\beta\cos\left(36°\right) = 1.618$$

There is one unpaired electron. The π energy is

$$E_\pi = 2\left[\alpha + 2\beta\right] + 3\left(\alpha + 0.618\beta\right)$$
$$E_\pi = 5\alpha + 5.85\beta$$

P13.21) Use the VSEPR method to predict the structures of the following:

a. PF_3 b. CO_2 c. BrF_5 d. SO_3^{2-}

a) PF_3 has 3 ligands and a lone pair. VSEPR predicts that the structure is pyramidal.

b) CO_2 has 2 double-bonded ligands and no lone pairs. VSEPR predicts that the structure is linear.

c) BrF_5 has 5 ligands and a lone pair. VSEPR predicts that the structure is square pyramidal.

d) SO_3^{2-} has 3 ligands and a lone pair. VSEPR predicts that the structure is pyramidal.

Computational problems

Before solving the computational problems, it is recommended that students work through Tutorials 1–3 under the Help menus in Spartan Student Edition to gain familiarity with the program.

Computational Problem 13.1: Calculate the bond angles in NH_3 and in NF_3 using the density functional method with the B3LYP functional and the 6-31G* basis set. Compare your result with literature values. Do your results agree with the predictions of the VSEPR model and Bent's rule?

Step 1: Create a new file, and build the NH_3 molecule using **nonplanar trivalent N** inorganic palette which is accessed by clicking the "Inorganic" tab on the model kit.

Step 2: Using the calculations menu, enter equilibrium geometry, the B3LYP method, the 6-31G* basis set and the appropriate number of unpaired electrons for the ground state of NH_3. Check the "infrared spectra" box. Click "OK."

Step 3: Using the "New Molecule" command and the same basis set and method, set up calculations for NF_3. Submit the calculation.

Step 4: Look at the vibrational frequencies by going to "display" and then "spectra." The result of the calculation is an equilibrium structure only if all frequencies are real.

Step 5: Using the measure angle button, record the H-N-H and F-N-F bond angles and compare your results with the predictions of the VSEPR model, Bent's rule, and with literature values.

Computational Problem 13.2: Calculate the bond angles in H_2O and in H_2S using the density functional method with the B3LYP functional and the 6-31G* basis set. Compare your result with literature values. Do your results agree with the predictions of the VSEPR model and Bent's rule?

Step 1: Create a new file, and build the H_2O molecule using **nonlinear divalent O** in inorganic palette which is accessed by clicking the "Inorganic" tab on the model kit.

Step 2: Using the calculations menu, enter equilibrium geometry, the B3LYP method, the 6-31G* basis set and the appropriate number of unpaired electrons for the ground state of H_2O. Check the "infrared spectra" box. Click "OK."

Step 3: Using the "New Molecule" command and the same basis set and method, set up calculations for H_2S. Submit the calculation.

Step 4: Look at the vibrational frequencies by going to "display" and then "spectra." The result of the calculation is an equilibrium structure only if all frequencies are real.

Step 5: Using the measure angle button, record the H-O-H and H-S-H bond angles and compare your results with the predictions of the VSEPR model, Bent's rule, and with literature values.

Computational Problem 13.3: Calculate the bond angle in ClO_2 using the density functional method with the B3LYP functional and the 6-31G* basis set. Compare your result with literature values. Does your result agree with the predictions of the VSEPR model?

Step 1: Create a new file, and build the ClO_2 molecule using **nonlinear divalent Cl** in the inorganic palette which is accessed by clicking the "Inorganic" tab on the model kit.

Step 2: Using the calculations menu, enter equilibrium geometry, the B3LYP method, the 6-31G* basis set and the appropriate number of unpaired electrons for the ground state of ClO_2. Check the "infrared spectra" box. Click "OK."

Step 3: Look at the vibrational frequencies by going to "display" and then "spectra." The result of the calculation is an equilibrium structure only if all frequencies are real.

Step 5: Using the measure angle button, record the O-Cl-O bond angle and compare your results with the predictions of the VSEPR model, Bent's rule, and with literature values.

Computational Problem 13.4: SiF_4 has four ligands and one lone pair on the central S atom. Which of the following structures do you expect to be the equilibrium form based on a calculation using the density functional method with the B3LYP functional and the 6-31G* basis set? In the figure below, from top to bottom, the structures are a trigonal bipyramid, a see-saw structure, and a square planar structure.

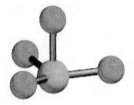

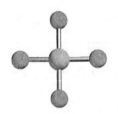

Step 1: Create a new file, and build trigonal bipyramidal SiF$_4$ using the appropriate S template in the inorganic palette which is accessed by clicking the "Inorganic" tab on the model kit.

Step 2: Using the calculations menu, enter equilibrium geometry, the B3LYP method, the 6-31G* basis set and the appropriate number of unpaired electrons for the ground state of ClO$_2$. Check the "infrared spectra" box. Check the "global calculations" box. Click "OK"

Step 3: Repeat step 1 for see-saw SiF$_4$ and square planar SiF$_4$.

Step 5: Using your results for the total energy of the different isomers, determine the equilibrium structure. Measure the 2 different F-S-F bond angle in your equilibrium structure and compare them with the experimentally determined values of 101.6° and 173°. Justify this structure by making a Lewis diagram for the molecule. Is this the structure that would have been predicted using the VSEPR model?

Computational Problem 13.5: Calculate the bond angles in singlet BeH$_2$, doublet NH$_2$, and doublet BH$_2$ using the Hartree–Fock method and the 6-31G* basis set. Explain your results using the Walsh diagram of Figure 13.11.

Step 1: Create a new file, and build the BeH_2 molecule using **nonlinear divalent Be** from the inorganic palette which is accessed by clicking the "Inorganic" tab on the model kit.

Step 2: Using the calculations menu, enter the equilibrium geometry, the Hartree–Fock method, the 6-31G* basis set and the appropriate number of unpaired electrons for the ground state of BeH_2. Check the "infrared spectra" box. The global calculations box should not be checked. Click "OK."

Step 3: Using the "New Molecule" command and the same basis set and method, set up calculations for doublet NH_2, and doublet BH_2. Submit the calculation.

Step 4: Look at the vibrational frequencies by going to "display" and then "spectra." The result of the calculation is an equilibrium structure only if all frequencies are real.

Step 5: Using the measure angle button, record the H-Be-H, H-B-H, and H-N-H bond angles and compare your results with the prediction of the Walsh diagram of Figure 13.11.

Computational Problem 13.6: Calculate the bond angle in singlet LiH_2^+ using the Hartree–Fock method and the 6-31G* basis set. Can you explain your results using the Walsh diagram of Figure 13.11? (Hint: Determine the calculated bond lengths in the molecular ion.)

Step 1: Create a new file, and build the LiH_2^+ molecular ion using **nonlinear divalent Be** from the inorganic palette which is accessed by clicking the "Inorganic" tab on the model kit.

Step 2: Using the calculations menu, enter equilibrium geometry, the Hartree-Fock method, the 6-31G* basis set and the appropriate number of unpaired electrons for the ground state of LiH_2^+. Check the "infrared spectra" box. Click "OK."

Step 3: Using the "New Molecule" command and the same basis set and method, set up calculations for the ground state of H_2, Submit the calculation.

Step 4: Look at the vibrational frequencies by going to "display" and then "spectra." The result of the calculation is an equilibrium structure only if all frequencies are real. Is LiH_2^+ a stable molecular ion?

Step 5: Compare the H-H distances in H_2, and the H-H vibrational stretching frequency. On the basis of these results, can you explain your results using the Walsh diagram of Figure 13.11?

Computational Problem 13.7: Calculate the bond angle in singlet and triplet CH_2 and doublet CH_2^+ using the Hartree–Fock method and the 6-31G* basis set. Can you explain your results using the Walsh diagram of Figure 13.11?

Step 1: Create a new file, and build the CH_2 molecule using **nonlinear divalent C** from the inorganic palette which is accessed by clicking the "Inorganic" tab on the model kit.

Step 2: Using the calculations menu, enter equilibrium geometry, the Hartree–Fock method, the 6-31G* basis set and the appropriate number of unpaired electrons for the ground state of CH_2. Check the "infrared spectra" box. The global calculations box should not be checked. Click "OK."

Step 3: Using the "New Molecule" command and the same basis set and method, set up calculations for triplet CH_2 and doublet CH_2^+. Submit the calculation.

Step 4: Look at the vibrational frequencies by going to "display" and then "spectra." The result of the calculation is an equilibrium structure only if all frequencies are real. Are all molecules stable?

Step 5: Compare the bond angles in the three molecules. Can you explain your results using the Walsh diagram of Figure 13.11?

Computational Problem 13.8: Calculate the bond angle in singlet NH_2^+, doublet NH_2, and singlet NH_2^- using the Hartree–Fock method and the 6-31G* basis set. Can you explain your results using the Walsh diagram of Figure 13.11?

Step 1: Create a new file, and build the NH_2^+ molecule using **nonlinear divalent N** from the inorganic palette which is accessed by clicking the "Inorganic" tab on the model kit.

Step 2: Using the calculations menu, enter equilibrium geometry, the Hartree–Fock method, the 6-31G* basis set and the appropriate number of unpaired electrons for the ground state of NH_2^+. Check the "infrared spectra" box. The global calculations box should not be checked. Click "OK."

Step 3: Using the "New Molecule" command and the same basis set and method, set up calculations for doublet NH_2, and singlet NH_2^-. Submit the calculation.

Step 4: Look at the vibrational frequencies by going to "display" and then "spectra." The result of the calculation is an equilibrium structure only if all frequencies are real. Are all molecules stable?

Step 5: Compare the bond angles in the three molecules. Can you explain your results using the Walsh diagram of Figure 13.11?

Computational Problem 13.9: How essential is coplanarity to conjugation? Answer this question by calculating the total energy of 1, 3 butadiene using the Hartree–Fock method and the 6-31G* basis set for dihedral angles of 0, 45, and 90 degrees.

Step 1: Create a new file, and build the 1,3-butadiene using sp^2 C from the organic palette which is accessed by clicking the "organic" tab on the model kit.

Step 2: Using the calculations menu, enter equilibrium geometry, the Hartree–Fock method, the 6-31G* basis set and the appropriate number of unpaired electrons for the ground state of 1,3-butadiene. Check the "infrared spectra" box. Click "OK."

Step 3: Using the "New Molecule" command and the same basis set and method, build 1,3-butadiene again. Click on "measure dihedral angle" button and then the 4 carbon atoms. Set this dihedral angle to 135° and hit Enter. You will now constrain this angle. From the "geometry" menu, choose "constrain dihedral" and click on the 4 carbon atoms. Click on the pink lock at the bottom corner of the window and enter 135°. Click minimize.

Step 4: Repeat the procedure of the previous step for a dihedral angle of 90°.

Step 5: Repeat the procedure of the step 3, this time constraining the H-C=C-H dihedral angle where the Cs refer to C1 and C2 to 135°. Click on "measure dihedral angle" button and then the H-C-C-H atoms using the trans H. Submit the calculation.

Step 6: Calculate the energy difference in kJ mol^{-1} between planar 1,3-butadiene and the three twisted structures. Which costs more energy, to rotate by 45° around the C1-C2 or the C2-C3 bond? Why?

Computational Problem 13.10: Calculate the equilibrium structures for singlet and triplet formaldehyde using the density functional method with the B3LYP functional and the 6-311+G** basis set. Choose (a) planar and (b) pyramidal starting geometries. Calculate vibrational frequencies for both starting geometries. Are any of the frequencies imaginary? Explain your results.

Step 1: Create a new file, and build the planar formaldehyde using sp^2 C from the organic palette which is accessed by clicking the "organic" tab on the model kit.

Step 2: Using the calculations menu, enter equilibrium geometry, the B3LYP method, the 6-311+G** basis set and the appropriate number of unpaired electrons for singlet formaldehyde. Check the "infrared spectra" box. The global calculations box should not be checked. Click "OK."

Step 3: Using the "New Molecule" command, sp^2 C from the organic palette and the same basis set and method, build planar triplet formaldehyde. Check the "infrared spectra" box. The global calculations box should not be checked. Click "OK."

Step 4: Using the "New Molecule" command, tetrahedral C from the organic palette and the same basis set and method, build nonplanar singlet and triplet formaldehyde. Delete one of the C valences. Check the "infrared spectra" box. The global calculations box should not be checked. Click "submit."

Step 5: Compare the geometry and total energy of the calculated singlet and triplet structures from the two starting geometries. Are they identical? If they are different, decide which structure is the true equilibrium structure, and explain why the two starting geometries gave different results.

Computational Problem 13.11: Calculate the equilibrium structure for Cl_2O using the density functional method with the B3LYP functional and the 6-31G* basis set. Obtain an infrared spectrum and activate the normal modes. What are the frequencies corresponding to the symmetric stretch, the asymmetric stretch, and the bending modes?

Step 1: Create a new file, and build Cl-O-Cl using nonlinear divalent O from the inorganic palette which is accessed by clicking the "inorganic" tab on the model kit.

Step 2: Using the calculations menu, enter equilibrium geometry, the B3LYP method, the 6-31G* basis set and the appropriate number of unpaired electrons for the molecule. Check the "infrared spectra" box. The global calculations box can be checked. Click "OK."

Step 3: Using the "New Molecule" command, build O-Cl-Cl using nonlinear divalent O from the inorganic palette which is accessed by clicking the "inorganic" tab on the model kit. Check the "infrared spectra" box. The global calculations box can be checked. Click "OK."

Step 4: Which of the two isomers has the lower energy? Compare the bond angle with what would be predicted by the VSEPR model.

Step 5: Obtain an infrared spectrum and activate the normal modes. Determine the frequencies corresponding to the symmetric stretch, the asymmetric stretch, and the bending modes. Compare these frequencies with experimental values from the chemical literature.

Computational Problem 13.12: Calculate the equilibrium structures for PF_3 using the density functional method with the B3LYP functional and the 6-31G* basis set. Obtain an infrared spectrum and activate the normal modes. What are the frequencies corresponding to the symmetric stretch, the symmetric deformation, the degenerate stretch, and the degenerate deformation modes?

Step 1: Create a new file, and build PF_3 using nonplanar trivalent P from the inorganic palette which is accessed by clicking the "inorganic" tab on the model kit.

Step 2: Using the calculations menu, enter equilibrium geometry, the B3LYP method, the 6-31G* basis set and the appropriate number of unpaired electrons for the molecule. Check the "infrared spectra" box. The global calculations box can be checked. Click "OK."

Step 3: Obtain an infrared spectrum and activate the normal modes. Determine the frequencies corresponding to the symmetric stretch, the symmetric deformation, the degenerate stretch, and the degenerate deformation modes. Compare these frequencies with experimental values from the chemical literature.

Computational Problem 13.13: Calculate the equilibrium structures for C_2H_2 using the density functional method with the B3LYP functional and the 6-31G* basis set. Obtain an infrared spectrum and activate the normal modes. What are the frequencies corresponding to the symmetric stretch, the antisymmetric stretch, the stretch, and the two bending modes?

Step 1: Create a new file, and build H-C≡C-H using *sp* C from the inorganic palette which is accessed by clicking the "organic" tab on the model kit.

Step 2: To avoid biasing the outcome of the calculation, make the molecule bent. Click on the measure angle button, and then on the three atoms. Set the angle to 150°. Hit the enter key. Do not minimize because the process will make the molecule linear.

Step 3: Using the calculations menu, enter equilibrium geometry, the B3LYP method, the 6-31G* basis set and the appropriate number of unpaired electrons for the molecule. Check the "infrared spectra" box. The global calculations box can be checked. Click "OK."

Step 4: Obtain an infrared spectrum and activate the normal modes. Determine the frequencies corresponding to the antisymmetric bending, symmetric bending, C≡C stretch, antisymmetric C-H stretch, and symmetric C-H stretch. Compare these frequencies with experimental values from the chemical literature.

Computational Problem 13.14: Calculate the structure of $N \equiv C - Cl$ using the density functional method with the B3LYP functional and the 6-31G* basis set. Which is more electronegative, the Cl or the cyanide group? What result of the calculation did you use to answer this question?

Step 1: Create a new file, and build linear $N \equiv C - Cl$ using *sp* C and N from the organic palette which is accessed by clicking the "organic" tab on the model kit.

Step 2: To avoid biasing the outcome of the calculation, make the molecule bent. Click on the measure angle button, and then on the three atoms. Set the angle to 150°. Hit the enter key. Do not minimize because the process will make the molecule linear.

Step 3: Using the calculations menu, enter equilibrium geometry, the B3LYP method, the 6-31G* basis set and the appropriate number of unpaired electrons for the molecule. Check the "infrared spectra" box. Click "submit."

Step 4: Using the output, decide if the Cl or the cyanide group is more electronegative.

Chapter 16: Molecular Symmetry

P16.2) Use the 3×3 matrices for the C_{2v} group in Equation (16.2) to verify the group multiplication table for the following successive operations:

a. $\hat{\sigma}_v \hat{\sigma}'_v$ b. $\hat{\sigma}_v \hat{C}_2$ c. $\hat{C}_2 \hat{C}_2$

$$\hat{E} = \begin{pmatrix} 1 & 0 & 0 \\ 0 & 1 & 0 \\ 0 & 0 & 1 \end{pmatrix} \quad \hat{C}_2 = \begin{pmatrix} -1 & 0 & 0 \\ 0 & -1 & 0 \\ 0 & 0 & 1 \end{pmatrix}$$

$$\hat{\sigma}_v = \begin{pmatrix} 1 & 0 & 0 \\ 0 & -1 & 0 \\ 0 & 0 & 1 \end{pmatrix} \quad \hat{\sigma}'_v = \begin{pmatrix} -1 & 0 & 0 \\ 0 & 1 & 0 \\ 0 & 0 & 1 \end{pmatrix}$$

a)

$$\hat{\sigma}_v \hat{\sigma}'_v = \begin{pmatrix} 1 & 0 & 0 \\ 0 & -1 & 0 \\ 0 & 0 & 1 \end{pmatrix}\begin{pmatrix} -1 & 0 & 0 \\ 0 & 1 & 0 \\ 0 & 0 & 1 \end{pmatrix} = \begin{pmatrix} -1 & 0 & 0 \\ 0 & -1 & 0 \\ 0 & 0 & 1 \end{pmatrix}$$

$$\hat{\sigma}_v \hat{\sigma}'_v = \hat{C}_2$$

b)

$$\hat{\sigma}_v \hat{C}_2 = \begin{pmatrix} 1 & 0 & 0 \\ 0 & -1 & 0 \\ 0 & 0 & 1 \end{pmatrix}\begin{pmatrix} -1 & 0 & 0 \\ 0 & -1 & 0 \\ 0 & 0 & 1 \end{pmatrix} = \begin{pmatrix} -1 & 0 & 0 \\ 0 & 1 & 0 \\ 0 & 0 & 1 \end{pmatrix}$$

$$\hat{\sigma}_v \hat{C}_2 = \hat{\sigma}'_v$$

c)

$$\hat{C}_2 \hat{C}_2 = \begin{pmatrix} -1 & 0 & 0 \\ 0 & -1 & 0 \\ 0 & 0 & 1 \end{pmatrix}\begin{pmatrix} -1 & 0 & 0 \\ 0 & -1 & 0 \\ 0 & 0 & 1 \end{pmatrix} = \begin{pmatrix} 1 & 0 & 0 \\ 0 & 1 & 0 \\ 0 & 0 & 1 \end{pmatrix}$$

$$\hat{C}_2 \hat{C}_2 = \hat{E}$$

P16.4) The D_3 group has the following classes: E, $2C_3$, and $3C_2$. How many irreducible representations does this group have and what is the dimensionality of each?

The D_3 group has three classes and six elements ($1E$, $2C_3$, $3C_2$).

There are three irreducible representations because there are three classes. Their dimensions are determined by the equation

$$d_1^2 + d_2^2 + d_3^2 = 6$$

This can only be satisfied by $d_1 = 2$, $d_2 = 1$, $d_3 = 1$. Thus

3 irreducible representations; two 1-dimensional and one 2-dimensional.

P16.7) XeF$_4$ belongs to the D_{4h} point group with the following symmetry elements: E, C_4, C_4^2, C_2, C_2', C_2'', i, S_4, S_4^2, $\sigma, 2\sigma'$, and $2\sigma''$. Make a drawing similar to Figure 16.1 showing these elements. The 4-fold axis is perpendicular to the plane of the molecule.

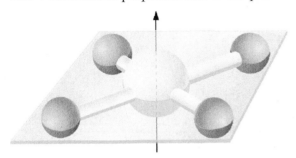

There are three 2-fold axes, as shown below. Two pass through the opposed F atoms, and the third bisects the F-Xe-F angle.

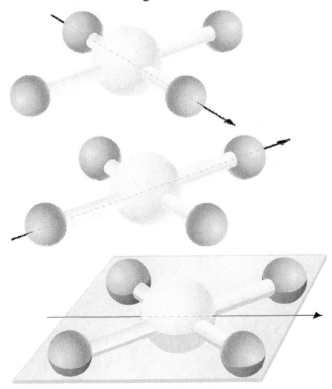

The mirror plane denoted σ lies in the plane of the molecule.

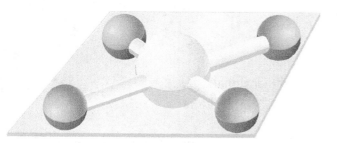

The two mirror planes denoted σ' are mutually perpendicular, and are perpendicular to σ. The Xe atom and two opposed F atoms lie at the intersection of σ' with σ.

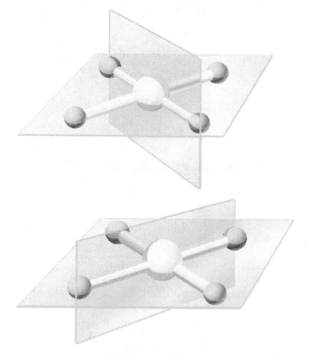

The two mutually perpendicular mirror planes σ'' contain the Xe, but no F atoms.

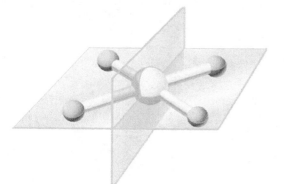

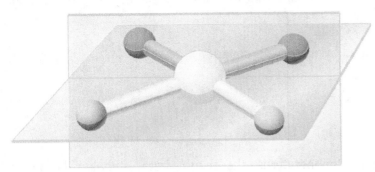

The Xe atom is the invasion center. This element is not depicted.

P16.13) Decompose the following reducible representation into irreducible representations of the C_{2v} group:

E	C_2	σ_v	σ_v'
4	0	0	0

We set up the following table and apply Equation (16.22). The reducible representation is in the first row.

	E	C_2	σ_v	σ_v'
Red	4	0	0	0
A_1	1	1	1	1
A_2	1	1	-1	-1
B_1	1	-1	1	-1
B_2	1	-1	-1	1

$$n_{A_1} = \frac{1}{4}\left[(1 \times 4 \times 1) + (1 \times 0 \times 1) + (1 \times 0 \times 1) + (1 \times 0 \times 1)\right] = 1$$

$$n_{A_2} = \frac{1}{4}\left[(1 \times 4 \times 1) + (1 \times 0 \times 1) + (1 \times 0 \times (-1)) + (1 \times 0 \times (-1))\right] = 1$$

$$n_{B_1} = \frac{1}{4}\left[(1 \times 4 \times 1) + (1 \times 0 \times (-1)) + (1 \times 0 \times 1) + (1 \times 0 \times (-1))\right] = 1$$

$$n_{B_2} = \frac{1}{4}\left[(1 \times 4 \times 1) + (1 \times 0 \times (-1)) + (1 \times 0 \times (-1)) + (1 \times 0 \times 1)\right] = 1$$

Thus: $\Gamma_{\text{red}} = A_1 + A_2 + B_1 + B_2$

P16.14) Show that z is a basis for the A_1 representation and that R_z is a basis for the A_2 representation of the C_{3v} group.

All of the operations of the C_{3v} group are described by the matrix equation

$$\begin{pmatrix} x' \\ y' \\ z' \end{pmatrix} = \begin{pmatrix} \pm 1 & 0 & 0 \\ 0 & \pm 1 & 0 \\ 0 & 0 & 1 \end{pmatrix} \begin{pmatrix} x \\ y \\ z \end{pmatrix} = \begin{pmatrix} \pm x \\ \pm y \\ z \end{pmatrix}$$

The signs of at least one of x and y are changed by the operations other than E, but it is always the case that $z' = z$. Therefore, the character of each operation on z is 1, which makes z a basis for the A_1 representation.

For a rotation about the z axis, the character is $+1$ if the direction of rotation is unchanged, and -1 if the direction of the rotation is changed. Because the identity operator does not change the direction of the rotation, the character is $+1$. The same is true of the $\hat{C}_2$ operator. However, reflection if both the σ_v and σ_v' planes changes the direction of rotation because the rotation axis lies in the plane. Therefore the character for these operations is -1. The sequence of these characters correspond to the A_2 representation.

P16.15) Use the logic diagram of Figure 16.2 to determine the point group for PCl_5. Indicate your decision-making process as was done in the text for NH_3.

1) linear?	No
2) C_n axis?	Yes C_3 axis $\Rightarrow z$ axis
3) more than 1 C_n axis	Yes
3) more than 1 C_n axis, $n > 2$?	No
4) σ plane?	Yes
5) $\sigma \perp$ to C_3?	Yes

We conclude that the point group is D_{3h}.

P16.20) Show that the presence of a C_2 axis and a mirror plane perpendicular to the rotation axis imply the presence of a center of inversion.

Under a 180° rotation around the z axis, $x, y, z \rightarrow -x, -y, z$. The effect of the mirror plane perpendicular

to the rotation axis is $-x, -y, z \rightarrow -x, -y, -z$, which is the same as the inversion operation.